Robotics

Everything You Need to Know About Robotics from Beginner to Expert

Peter McKinnon

Peter McKinnon

© Copyright 2015 - All rights reserved.

In no way is it legal to reproduce, duplicate, or transmit any part of this document in either electronic means or in printed format. Recording of this publication is strictly prohibited and any storage of this document is not allowed unless with written permission from the publisher. All rights reserved.

The information provided herein is stated to be truthful and consistent, in that any liability, in terms of inattention or otherwise, by any usage or abuse of any policies, processes, or directions contained within is the solitary and utter responsibility of the recipient reader. Under no circumstances will any legal responsibility or blame be held against the publisher for any reparation, damages, or monetary loss due to the information herein, either directly or indirectly.

Respective authors own all copyrights not held by the publisher.

Legal Notice:

Peter McKinnon

This book is copyright protected. This is only for personal use. You cannot amend, distribute, sell, use, quote or paraphrase any part or the content within this book without the consent of the author or copyright owner. Legal action will be pursued if this is breached.

Disclaimer Notice:

Please note the information contained within this document is for educational and entertainment purposes only. Every attempt has been made to provide accurate, up-to-date and reliable complete information. No warranties of any kind are expressed or implied. Readers acknowledge that the author is not engaging in the rendering of legal, financial, medical or professional advice.

By reading this document, the reader agrees that under no circumstances are we responsible for any losses, direct or indirect, which are incurred as a result of the use of information contained within this document, including, but not limited to, —errors, omissions, or inaccuracies.

Contents

Introduction 7

Chapter 1 Robots 9

Chapter 2 Hardware Tutorial 17

Chapter 3 An Introduction to RoboCORE 31

Chapter 4 Software Tutorial 39

Chapter 5 Materials for Building a Robot 57

Chapter 6 Tips 67

Chapter 7 Circuit Schematics 79

Chapter 8 How to Build a Simple Robot for Beginners 89

Chapter 9 A Simple Touchless Sensor Robot 101

Chapter 10 Making an Autonomous Wall-Climbing Robot 141

Chapter 11 Cognitive Robotics 161

Chapter 12 Cloud Robotics 167

Chapter 13 Autonomous Robotics 179

Chapter 14 Different Types of Robots 187

Chapter 15 A Deeper Dive into Robotics 195

Chapter 16 How to Go About Using Motor Controllers and Microcontrollers 205

Chapter 17 How to Control Your Robot Through the Use of Sensors 213

Chapter 18 How to Assemble Your Robot and Program It Correctly 221

Conclusion... 237

Introduction

When we talk about robots, many people picture a machine imitating a human, like the robots from "Terminator" or "Star Wars." These are the ones found in science fiction movies, robots with consciousness. However, robots with such superior thinking ability don't exist in reality. Though we have successfully given enough common sense to robots for interacting with this dynamic world, we couldn't give consciousness to a robot. Giving common sense is one thing but giving consciousness is on a whole different level. Many industries and scientists are trying to create these types of humanoid robots.

The type of robots that we encounter frequently are those designed to do tasks that are too onerous, boring, dangerous, or just plain nasty for humans. Robots of this kind are being widely used in the areas of manufacturing, automobiles, medicine, and space. Robots have become a part of our lives directly and indirectly. In this book we will discuss the second type of robots.

To understand the concept of robotics, one needs decent knowledge in the areas of electronics, mechanics, and programming. With the increasing use of robots in our day-to-day lives, one can actually find a good career in robotics. You can start learning robotics by making simple robots. These are actually fun to use.

I hope you find this book useful and I thank you for using this book.

Chapter 1

Robots

In the year 1979, the Robot Institute of America defined a robot as

"A reprogrammable, multifunctional manipulator designed to move materials, parts, tools, or specialized devices through various programmed motions for the performance of a variety of tasks."

Webster defined a robot in a more inspiring way: "An automatic device that performs functions normally ascribed to humans or a machine in the form of a human."

Origin of the Word 'Robot'

Karel Capek (1890-1938), a Czech playwright, used the word "robot" for the first time, deriving it from a Czech word meaning serf or forced labor. He was very prolific and influential as a playwright and a writer. Because of his works, he was considered a candidate for the Nobel Prize, reportedly several times. He used the word robot for the first time in his play *R.U.R.* in the year 1921. "R.U.R. stands for "Rossum's Universal Robots" and the play is about robots and the advantages they bring. It ends ironically with blight in the form of social unrest and unemployment. The theme of this play, in part, was to show the dehumanization of man living in a technological civilization.

You may be surprised to hear this but Capek's robots were not mechanical. They were created through chemical means. Capek said that he thought this idea was possible in an essay published in 1935.

We can say that Capek's brother Josef, also a writer, actually coined the word robot. The evidence is that Capek asked Josef, in a short letter, about the name to be given to the artificial workers in his play.

Capek suggested the word Labori but thought that this word was too 'bookish.' His brother, Josef muttered, "Call them robots" and went back to his work. And so, we have the word robot from a curt response.

First Use of Word 'Robotics'

Isaac Asimov used the word "robotics" in his short story "Runaround," for the first time in the year 1942. A collection of stories, *I, Robot*, was published in 1950.

"Runaround" was one of Asimov's first works on robots. Eliza is the modern counterpart of his fictional character. A professor from MIT, Professor Joseph Weizenbaum, wrote Eliza in the year 1966. He described it as "a computer program for the study of natural language communication between man and machine."

She was designed to simulate a psychotherapist with her 240 lines of initial code, which enabled her to answer the questions with questions.

Three Laws of Robotics

In his work "I, Robot," Isaac Asimov proposed the three laws of robotics. Later, he added the zeroth law.

Law Zero: A robot may not injure humanity, or, through inaction, allow humanity to come to harm.

Law One: A robot may not injure a human being, or, through inaction, allow a human being to come to harm, unless this would violate a higher order law.

Law Two: A robot must obey orders given it by human beings, except where such orders would conflict with a higher order law.

Law Three: A robot must protect its own existence as long as such protection does not conflict with a higher order law.

Unimate - The First Robot

In 1956, there was a technology explosion. At a historic meeting between Joseph F. Engelberger, an engineer, and George C. Devol, a successful entrepreneur and inventor, the foundation was laid for the robotics industry. They discussed the works of Isaac Asimov.

They both made a commercial and serious decision to develop a real, working robot. They successfully persuaded Norman Schafler from the Condec Corporation that they could make it a commercial success.

Joseph F. Engelberger then started a manufacturing company called Unimation, which stood for universal animation. This was the first commercial company that made robots. George C. Devol took care of writing the necessary patents. They named their first robot "Unimate." As a result, Joseph F. Engelberger is considered the "father of robotics."

They installed their first Unimate at a GM plant for working along with the heated die casting machines. Most of these robots were sold to extract the die casting from machines and to perform spot welding on metal bodies. Both of these tasks were particularly hateful jobs.

The applications became a commercial success, with robots working reliably. They helped the industries that purchased them to save a lot of money as they replaced people. This was the beginning of a new industry and robots started performing a variety of other tasks, which included loading and unloading machine tools.

Westinghouse ultimately acquired Unimation and fulfilled the entrepreneur's dream of wealth. The company is still in production, selling robots.

People came to know about the potential of robots and the idea of robots was hyped to the skies. It became a high fashion, as many presidents from large corporations came forward and bought them for roughly $100,000 apiece. They got them and put them in their laboratories just to see what they were capable of. The sales have constituted a very large part in the robotics market. Many companies have reduced their return on investment criteria on robots for encouraging industries to use them.

Benefits of Robots

Robots provide specific benefits to countries, industries, and workers. If they are introduced correctly, the industrial robots can possibly improve the quality of life by relieving workers from dangerous, boring, dirty, and heavy labor. It is true that when a robot replaces a worker, it causes unemployment, but it also create jobs for programmers, engineers, salesmen, robotic technicians, and supervisors.

Improved productivity and management control and producing quality products consistently are some of the benefits offered by the robots to the manufacturing industry. The industrial robots can continuously work night and day tirelessly on an assembly line, without any change in their performance.

With this, they can reduce the cost of manufactured goods to a great extent. Countries using robots for industrial use will have an advantage economically.

Peter McKinnon

Chapter 2

Hardware Tutorial

Robot Brains

It is great fun to build a robot. Imagine robots that can think for themselves. It is not hard to add a thinking brain to your robot. This brain will help your robot follow rules and instructions. There are two basic types of robot brains, analog and digital.

Analog Brains

Using hardwired circuits, the actuators of the robot can be controlled. You can make your robots follow simple rules by designing circuits from transistors, capacitors, and resistors. For instance, if a robot hits an obstacle, a simple switch on the robot's body would be pressed to make the robot go back and turn, hopefully avoiding obstacles on its path.

Analog brains might look simple but they have disadvantages. For complex robots, it is extremely difficult to design an analog brain. You need a good knowledge of electronics for your design. They are extremely difficult to change after building. You might need to rebuild the analog brain of your robot completely, if you wish to change your robot's behavior.

For beginners in robotics or electronics, using analog circuits is not recommended. We have another option we can use - the digital brain.

Digital Brains

Small devices called microcontrollers are perfect as the brains for robots. In simple words, microcontrollers are nothing but small computers at the chip level. They have their own processor and memory. You can program them to control a robot any way you wish. You can connect them to a PC and can program them.

The best part is that they can be programmed any number of times just by clicking a mouse. With microcontrollers, you don't have to worry about using a soldering iron for components.

It's very easy to program the chips and to fully understand them; all you need is a bit of patience. Trying to learn programs using the textbook is a boring and slow process. It is difficult to memorize the programs from a textbook. Comparatively, it is far easier to program by trying out some examples from a tutorial. By playing with them and by applying your own ideas, you can get a better understanding of how programs work. This way is fun, too. When you are confident, you can start writing your own programs.

Motors

Motors are by far the most commonly used method for moving robots. You can add mobility to a robot by simply connecting a motor to its wheels or gears. There are many types of models available that can be used for robotics. In this tutorial, we will look at some of the most commonly used motors.

DC Motors

DC motors are the easiest and most commonly used motors for robots. They have two terminals that will be connected to their power supply. You can change the rotation of the motor by reversing its polarity. Reversing polarity is done by swapping the positive and negative terminals of the power supply.

Unfortunately, compared to processors, motors use a lot of power, so they cannot be connected directly to processors. The processor cannot supply enough current to run the motor. We will need an external source for turning the motor on and off. For this, we can use driver chips, relays, or transistors.

There are two motor-driven chips on the RoboCORE that can simultaneously control 4 DC motors. Connecting these motors is very simple. You are ready to go after connecting the two wires from each motor to one of the outputs of the RoboCORE motor. The output pins on the processor control the motor. Let us say that the two pins are pin1 and pin2.

The rotation of the motor can be changed by changing the outputs of the pins. It is given in the below table.

Pin 1 Pin 2 Motor Output

Pin 1	Pin 2	Motor Output
On	Off	Clockwise
Off	On	Anti-Clockwise
Off	Off	Motor Off

For help on programming the chip to do this, look at the motor programming guide.

We have learned that the direction of the motor can be changed by reversing its polarity, but that is not the only way of controlling the motor. The speed of the motor can be varied by changing the voltage given to the motor. But these motors can only have two settings, either on or off. So, how can you temporarily change the voltage of the motor? The answer is a technique called "pulse width modulation."

Pulse Width Modulation

This is a technique in which the motor is fed with pulses of electricity, producing an average voltage effect by giving the pulses at a fairly fast rate. We will look at a few examples that will help you understand this well.

Let us say that we turn on 20 volts for 80ms (80 thousandths of a second) and turn it off for 20ms. If we continue to do this over and over, you're basically changing the voltage so quickly that on and off will become an average voltage. In this case, we are turning off the voltage for 1/5 of the time (20% of the time). The average voltage given to the motor will be 80% of 20 volts, which is 16 volts. Instead of running at full capacity, this will slow down the motor. With this method, you can change the motor speed by changing the on and off time of the motor and by turning the current off.

We can use the command DACPin with the motor drivers present on the RoboCORE. You can find the full syntax of the DACPin command from the document files given with the BasicX software. The basic command line is given below:

Call DACpin (Pin, Voltage, DACcounter)

Pin = The output pin

Voltage = Byte value between 0 and 255

DACcounter = A value must be returned to this variable by the function. In case of multiple pins using the DACPin command, each of those pins should have different variables.

Depending on the voltage, the results obtained after running the pulse through the motor-driven chips will vary. Typically, a reduction of 25% in power can be achieved. If the voltage is further reduced, most DC motors may stop working due to insufficient power supply.

Torque

Torque can be defined as the turning force applied on a body. The torque is a measurement for the power of a motor. If a motor has higher torque, it can move greater weights. Varying amounts of torque are produced by DC motors, depending on the speed they are running. The running speed is measured in revolutions per minute or

RPM. DC motors produce less torque at lower RPMs and generally produce higher torque with higher RPMs.

For instance, let us say that a robot is powered by 2 DC motors. Reducing the overall speed by using gears and while running them at their top speed will deliver the most power to the wheels. You can see that the motors deliver the required amount of torque by using pulse width modulation for slowing your motor.

Pulse width modulation is an important and useful technique, as it allows you to control the speed of the robot, using software alone. In some situations, you might need to slow your robot down to move away from obstacles or to change its direction.

Servo Motors

Servo motors, or simply servos, are perfect for controlling. You can specify a position for them to rotate. Servo motors are ideal for anything that includes precise moment. Most of these can rotate from 90° to 180°. Some servo motors can go all the way to 360°. However, they cannot continually rotate and for this reason they cannot be used for powering wheels. Their precision moment is perfect for controlling rack and pinion style steering, powering legs, and much more.

Servos are completely self-contained. They include a gearbox, motor, and driver electronics, making it possible to control them directly from a microcontroller, without using interface electronics.

Servo motors have three wires, two of which are for supplying power, usually anywhere between 5 and 7 V. The third wire is for controlling and can be directly connected to the microcontroller or processor. The position to which they rotate can be given to them by sending electric pulses to the servo. You can directly control the position by changing the delay between electric pulses.

How Servos Work

For you to understand how a servo motor is controlled, you should take a closer look at how they work. There is a set of gears, control board, a potentiometer, and a motor inside a servo. The potentiometer is nothing but a variable resistor; a gear set connects it to the motor. The position of the motor is given by the control signal and, based on that, the motor turns. Along with the motor, the potentiometer also rotates, changing its resistance. The control circuit will monitor this resistance and once it reaches the appropriate value, it will stop the motor. This means that the servo has reached the correct position.

Controlling Servos

Using pulse width modulation, servos are positioned. With this method, the server is sent a continuous stream of electric pulses. Depending on the servo positioning, the electric pulses usually last for 1 ms or 2 ms. To keep the servo in its position, continuous electric pulses should be sent. Usually, 50 to 60 electric pulses a second will be sent. The position of the servo depends on the pulse and not on the number of times it is sent.

A 1 ms pulse will set the servo position to 0° and the 2 ms pulse will set the servo position to the maximum that it can rotate. By sending a pulse of 1.5 ms, the position of the servo will be halfway around its complete rotation.

Stepper Motors

Stepper motors are similar to DC motors in the way they work. The difference is that the DC motor has a single electromagnetic coil for producing movement, whereas a stepper motor contains many of them. Stepper motors can be controlled by turning each of those coils on and off in a sequence. Whenever a new coil is turned on, the motor will rotate a few degrees; this is called the step angle. By repeating these sequences, the motor will continuously turn a few degrees with every coil energized, making it a complete and constant rotation. For example, if there is a stepper motor with 7.5° step angle, it will require 48 pulses to complete one revolution.

There is a magnet in the middle of the stepper motor arrangement; the motor shaft is connected to it. This produces the rotation. Four magnets are located on the outside and they represent each of the stepper motor's coils. The magnet in the center gets pulled in the direction of the energized coil. The motor can be rotated when the correct sequence of pulses are applied.

Sensors

We live in a complex world and we use our senses for understanding the surroundings. Like humans, robots also need senses to safely move around by understanding their surroundings. The easiest way to do this is by adding simple sensors.

Bump Sensor

So you have added some motors for your robot to run and it is driving around happily. It probably gets stuck when it collides with obstacles. You need a solution that detects obstacles and moves around them. For this, you can use something called a bump sensor, which will let your robot know that it collided with something. The simplest way is to add a micro-switch in front of the robot so that it gets pushed when your robot collides with something. This switch will send an electric signal and the microcontroller will change the direction of your robot. The switches are normally held open using an internal string.

Micro-switches are digital and they can be easily connected to a microcontroller. They can only have two values, on and off. All the microcontrollers are digital and they can be matched with micro-switches. They can be connected to any free socket.

The resistor holds the signal line when the switch is off. Without their assistance, the signal line will be "floating," as it is not connected to anything. The processor will constantly be trying to decide the status of the line and, without a resistor, the signal line might cause unreliable readings.

Light Sensor

If you wish to make your robot interesting, add a light sensor. Light sensors will make your robot follow light or hide in the dark. A light sensor is nothing but a resistance that changes value based on the amount of light it receives.

A light sensor can be easily connected to the free analog socket of the microcontroller, using a simple circuit. It is very easy to get results. Make the processor take the reading from the sensor. When it gives a low value, it means that there is a lot of light on the sensor and, if the sensor gives a high value, it indicates that there is not much light falling on it.

Bodywork

Now that you have learned how to use the basic electronics for controlling your robot and for giving it a thinking control station, you should learn to build a platform for holding all of these bits.

If you're not sure about the design, you can go online and search for new designs. You can use those designs intact or add some of your own ideas to them.

Getting into the garage is a great way to build a robot chassis. You can actually build a chassis from scrap but you should make sure that it is strong. Your robot design completely depends on your imagination and the things you have got. You can make your robot from a hard drive cage. You can even use a wooden plank as a base. Using LEGO bricks as your base is a very good option; it might not be as strong as metal, but it gives you the comfort of being able to dismantle your robot easily if you want to change its design. Once the base is set, it is not hard to find the electric motors. You can get them from old toy cars or from any other old devices with a motor. Just make sure that the motor is powerful enough to run your robot. After getting these, you need a power source for running the motors. You can use batteries as your power source. The batteries and electronics can be easily attached using plastic bags or elastic bands.

Chapter 3

An Introduction to RoboCORE

There are various methods for achieving the results you want out of a robot. Usually, building a robot would involve your schematic plan for execution, hardware, software, and the mechanics you can use to work the robot. With the introduction of RoboCORE, the manipulation of hardware and the software becomes much easier, allowing you to start with your own robot in just a few simple steps!

RoboCORE is the result of a successful crowdfunding application. It is a gadget and a development platform controlled by the cloud and it is designed to be the heart and the brain of your new robot. Using RoboCORE allows you to build your very own unique robot without the need for any experience at high-level programming and at a much lower cost than having to source your own motors.

RoboCORE gives you the right combination of programming and equipment needed to make just about any type of robot you can think of. It does not rely on any one specific mechanics framework so you can use any basic development, including metal and even LEGO® to build your robot.

In the same way, you can connect your USB camera or your smartphone to RoboCORE, further enhancing the abilities and characteristics of your robot. RoboCORE is not just the driver, it is a complete ecosystem, a system that allows you to use your imagination and pave the way for fantastic developments in the world of robotics.

RoboCORE is used to provide an ecosystem comprising a complete amalgamation of hardware and software solutions for programming and driving your robots or any peripheral connected to the robot. The system contains a RoboCORE physical unit, a cloud service, and the programming library.

RoboCORE has officially become a reality under a new name, Husarion. The hardware components of the RoboCORE project include a main processor containing a Cortex-M4 processing unit and also motor drivers, GPIO, ADC, and serial interfaces that are compatible with NXT motors and Mindstorm's EV3, which includes sensors and controller bricks.

Another option that is implemented is the integrated Intel Edison MCU that is capable of handling Bluetooth connectivity and Wi-Fi; this also extends a provision to attach a smartphone or a USB port for Internet access or even to a Raspberry Pi and other devices that are compatible.

One of the most interesting peripherals that are compatible with the RoboCORE is the CAN bus, a message-based protocol originally developed specifically for cars and in-car communications between multiple modules. It was later adopted for real-time distributed control in automation applications within industrial environments. Without significantly affecting the response time during execution, the CAN bus allows up to 20 RoboCORE units to be linked at a time.

On top of that, RoboCORE is also driven by a real-time operating system (RTOS) and it is therefore more suitable for applications that require a quick response, a regulator that is used in automaton tasks, or even for a root balanced on two wheels.

While RoboCORE has its fair share of uses in the vast field of robotics, it is integrated into the cloud server that helps you program and control a unit from any location at any instant in time; all you have to do is connect to it through a web browser or an iOS or Android smartphone application. With the help of these platforms and the powerful Hframework library, it allows high-level programming in C++ and Python. Once the project is successfully funded, the library will be made available as an open source.

Why Is RoboCORE the Best?

You only have to take a look at what goes into RoboCORE and the benefits it can provide to robot enthusiasts all over the world to see why it is the best solution:

- **Low-cost advanced hardware**. RoboCORE is an ARM Cortex M4 microcontroller, an Intel® Edson computer that operates on Linux. It has motor drivers built in, complete with an encoder interface. It includes extension ports and sensors, all rolled into one neat package that allows you to do whatever you want.

- **Powered by the cloud.** RoboCORE is fully powered by the cloud, allowing you to program and control your robotic creation from anywhere, as well as being able to share your construction project with anyone you want. All you need is a web browser and a good Internet connection.

- **Easy to program.** RoboCORE uses the Python or C++ programming interface and, if you have ever delved into programming the Arduino, you will have no trouble with the RoboCORE.

- **The software is open source.** This means that the software is open, anybody can use it, and anyone can modify it for his or her own use. It is fully hackable (in a good way) and fully extensible.

RoboCORE is the simplest way to build a robot whether it is for a consumer market or for your own personal use. It is also completely compatible with LEGO®

MINDSTORMS®, further increasing your opportunities to construct something quite unique.

Basic Specifications of RoboCORE

RoboCORE is often called the "heart" of a robot and is said to simplify the process of making your own robot. There are two types, the RoboCORE full-featured unit and the miniaturized version, which includes fewer sensor, motor, and expansion ports. This version does not support the CAN bus port but also features six internal servo ports. Here is a chart comparing the different types of RoboCORE.

RoboCORE	**RoboCORE mini**
Dimensions:	
4.53×4.92 inch/115x125 mm	3.2×3.2 inch/82x82 mm

STM32F4 168MHz ARM Cortex-M4, 196KB RAM + optional Intel Edison, 1 MB Flash Drive

Bluetooth 4.0 and 2.1 via Intel Edison, dual-band 802.11 a/b/g/n Wi-Fi

1 CAN bus port	–
1 USB host, 1 USB slave + 1 USB host via Intel Edison	
SD card interface	

hMotor DC motor + encoder input ports:

6	2
–	6 internal servo ports

hSensor I2C/UART, ADC, GPIO, interrupt channel Sensor ports:

5	4

Maximum external servo ports via expansion module:

24	12

hExt I2C/UART, ADC, GPIO, SPI, interrupt channel Expansion ports:

2   1

Power supply: 6-14 VDC

Husarion is a startup from Poland that initially came up with the idea of RoboCORE and currently comprises a team of highly skilled robotics enthusiasts and engineers who have various projects and research material on robotics in their portfolio.

Chapter 4

Software Tutorial

Introduction

This introduction will introduce you to programming the BasicX microcontroller. Don't worry if you are new to programming; this tutorial will help you. If you are an experienced reader, you are advised to skim through the topics to get the basics of the language.

In simple terms, programming is the common language for you and your RoboCORE. It instructs your RoboCORE what to do. Programming is required in robotics to make the machines work without human intervention, all by themselves. These robots are defined as autonomous robots.

1. We will start by connecting the serial port of your computer and power supply to your RoboCORE. We will also load the BasicX software.

2. After loading the BasicX software, select the COM port connected to your RoboCORE from the Monitor Port menu. The port is probably COM1. Set the Download Port menu to that port.

3. Click on the editor button and it will open a window asking for a file name. Give a name for your first program. After giving the name, it will tell you that there is no such file and it will ask you if you want to create it. Select "Yes."

Now we will write our first program.

First, we should make sure that the BasicX is correctly working and for this we will instruct it to send your computer a message. We will use the debug.print command for this.

Type or copy the code below to your editor window. Just enter the middle line in case the first line and last line are already present.

Sub main ()

debug.print "RoboCORE test: Everything is working"

End Sub

We should set the chip preferences before downloading the BasicX. This will basically specify the function of each pin on the chip. For doing this, select the project menu and click "Chip," or you can simply press the F7 button. For now, we will leave things as they are, so just click on "Okay."

Now click on the Compile menu and select "Compile and Run" (keyboard shortcut F5). This will send the code to the RoboCORE. Going back to the main BasicX window, you should see the text "RoboCORE test: Everything is working" appear on the screen.

The "Sub main ()" and "End Sub" commands tell the RoboCORE where the program begins and ends.

The debug.print command is very useful for telling us what the RoboCORE is doing; in the next case we will see more of its capabilities.

Now we will make the RoboCORE solve an equation for us. To do this we will need a variable. This is a value stored under a name in the memory of the RoboCORE. We will give this variable the name "answer."

Sub main()

Dim answer As integer ' declaring

Dim answer As string ' variables

answerI = 5*5 ' doing the math

answer=Cstr(answerI) 'convert to printable format

debug.print "5 times 5 is "; answer 'print answer

End Sub

Run the above program before starting and the computer should display the answer. If this succeeds, you know that your program can print values to the screen with the debug.print command, simply by adding a semicolon and by adding a variable name after 'text'. This program can be handy for testing sensors.

You can also use comments in this program and they will have no effect. Comments are very useful for adding information. The system will ignore any text written after ('). We have also used variables in this program. We will discuss about variables in the next topic.

Variables

Variables are nothing but values stored on the computer. Variables are needed for almost every program to help the RoboCORE remember things. There are various types of variables, each assigned to a specific data type. For example, text values are taken by strings; whole number values are taken by integers.

You need to specify the type of value that you intend to use or the computer will get confused.

Every variable has its own advantages and limitations; for this reason, we have simple methods that switch the variable types. Now we will have a look at the below program, line by line.

Sub main()
Dim answer As integer ' declaring
Dim answer As string ' variables

The variables are declared in the Dim lines. It means that they are set up to be used in the program. We use the 'As' keyword for specifying the type of the variable. The above simple program has a string and an integer.

answerI = 3*6 ' doing the math

Here, an integer variable is declared as it can be used for performing mathematical functions like multiplication.

answer=Cstr(answer) 'convert to printable format

The Cstr statement in the above line is used for converting the integer value stored in answer to a string value, which is a text variable. This can be printed on the screen using the debug.print command.

debug.print "3 times 6 is "; answer 'print answer
End Sub

Variables can be saved with any name you like, except for the words used in the programming language itself. It is advised that you keep the variable names simple, like assigning words that describe what the variable does or a single letter.

There are many other variable types available; one such is the Boolean variable. This is very commonly used in robotics. It can hold only two values, true or false.

For instance, there are only two possible outcomes with bump sensors, either be pressed or not. The Boolean value for the bump sensor will return "true" if pressed and "false" if not.

Boolean variables can be set using the below statement.

Dim variable As Boolean.
variable = true

Boolean logic is usually used with conditional statements; we will discuss these in the next chapter.

Another type of variable is the single variable type. These are used for storing float values like 3.84 or -1.47. In robotics, these use complex mathematical numbers.

Obstacle-Avoiding Robot

In this tutorial, we will make a fairly simple robot. We will use the Sharp GP2D02 IR, a proximity sensor, RoboCORE, and a few micro-switch bump sensors. A sample program for this is given later in this chapter.

This is a small robot working with infrared vision. The Sharp IR provides this vision up to a range of 80 cm. It is fun to watch the robot drive by "seeing" the objects in its path. It is not foolproof, unfortunately, and it might not detect small objects. So, adding bump sensors on its front is a needed backup.

You can use any chassis for a robot. Any metal, plastic, wood, or LEGO blocks can be used as the chassis for the robot. Using the LEGO as your chassis is a good idea, as it can be dismantled easily, in case you want to change your design. It is easy to deal with and modifications are simple. When preparing a chassis for your robot, make sure that you use a strong one, which can take a little weight and is strong enough to absorb accidental impact.

The micro-switch bump sensor should be added in the front. You can attach them by using glue or with the help of elastic bands. You can also use a sliding bump bar and masking tape with it. Use anything, but make sure that the micro-switch bump sensors stay intact. The elastic bands will help you to reset after a collision. These elastic bands can be added on either side of the chassis.

If you intend to use your robot on a track, it is better to use light parts and good gearing, so that the robot won't have problems when turning.

Now that the chassis for your robot is ready, we should place all the electronics on it. In this tutorial, we will use RoboCORE, a Sharp GP2D02 IR sensor, a Tamila to 2.1 mm converter. We will depend on the 9.6 V battery pack to supply the power.

You can solder the bump sensors to the 0.1" headers with 47K pull-down resistors.

The Sharp IR sensor is soldered to 0.1" pitch headers.

You should make the following connections to follow the sample program.

- The DC motor terminals should be connected to outputs 1 and 2.

- The bump sensors should be connected to pins 10 and 12.

- The sharp infrared sensor should be connected to pins 13 and 15.

To set up the robot, you need to put the battery in a convenient place on the RoboCORE. You can secure all the parts using elastic bands. On the front of your chassis, the Sharp IR sensor should be placed using elastic bands.

The robot will be driven by the software. It will move forward until it detects an object. The IR sensor will give an indication to the microcontroller if it meets an obstacle. The microcontroller will then power the servos and make the robot turn. If the obstacle is small, the Sharp sensor might fail to notice it and the robot will hit the obstacle. The bump sensor in the front will then come into play. When the robot bumps into an obstacle, the micro-switch of the bump sensor will be pressed and a signal will be sent to the microcontroller, which then turns the robot. You can put your robot in a corner and can have fun watching it dance.

Controlling DC Motors

We have already discussed the principles of DC motors in earlier sections. In this section, we will learn about the electronics that are needed to interface DC motors to the microcontroller or to any other digital chip.

Using RoboCORE is the easiest way to control the motors, as it has the driver electronics. These driver electronics can control up to two stepper motors or four DC motors at the same time. It also has direction LEDs that help us to debug the circuits easily.

However, if you plan on using your own motor drivers for your custom projects, you can use them with various methods.

You can use a circuit called H Bridge for controlling the direction of your DC motor.

Here's how the circuit works:

Small devices called transistors can be used to change the polarity of the power supply given to the motor. Transistors are nothing but small electronic switches that turn large voltages using small current. Every pair of transistors will be connected to a microcontroller pin. The requirement for the components to use will depend on the capacity of the motor that you are using.

The diodes play a very important role in the circuit. They are called fly-back diodes. They protect the transistors by preventing the voltage spikes coming from the motor. Whenever a motor rotates and changes its direction, the wire coils inside the motor act like a generator, producing current. This current generated by the motor is called back EMF (electromotive force). This current will travel through the wires back to the circuit as powerful voltage spikes heading toward the transistor. We already know that transistors allow the current to flow in a single direction. The transistor will stop the current flow from flowing toward the circuit, preventing it from blowing. If the back EMF is strong, it can possibly blow the transistor, giving your circuit a short life. So it is very important to use a transistor of the right capacity.

Protecting the transistor is the job of the diode. It will reroute the back EMF to the battery safely by bypassing the transistor. The power supply to the circuit can be any voltage suiting your motors. In cases where you are using two different power sources, one to cover your motors and the other to run your BasicX, you should connect each supply to the ground or the circuit might not work properly.

If you want to use specific components for your circuit, the following components will handle most of the motors up to 12 V.

Diodes: IN4002

Transistors: TIP41 NPN power transistor

Resistors: 2.2K ohm, 0.25 watt

Choosing the Right Motor

We use electric motors to actuate your robot's components, such as tracks, legs, wheels, sensor turrets, fingers, arms, or weapon systems. There are various types of electric motors but, when dealing with amateur robots, usually choice comes down to the following three:

- When power is applied to the continuous DC motor, it will continually rotate the shaft. The shaft of the motor will only stop when the power supply is removed or when the motor stalls if it cannot drive the attached load.

- Unlike a DC motor, the shaft of the stepper motor only rotates a few degrees and stops. For the stepper motor to rotate continuously, the power should be pulsed to the motor. The stepper motors are divided into three subtypes; one of them is the continuous DC motor. Most of the time we will be using the permanent magnet stepper motors. They are the easiest to use.

- Another type of motor is the servo motor, or servo for short. Servo motors are considered a special subset to the continuous DC motors. Stepper motors ensure positional accuracy by combining a feedback loop to the continuous DC motor. There are various types of servos and one of these types is found in hobby and model radio-controlled planes and cars.

It is hard to select the best motor for amateur robots from among DC motors, stepper motors, and servos. Now we will look at the pros and cons of the three motors and you can select the one you want based on this analysis.

Continuous DC Motor
Pros:

- A wide range of both new and used continuous DC motors are available.

- They can be easily controlled through a computer using electronic switches or relays.

- When combined with a perfect gearbox, larger DC motors are capable of powering 200-pound robots.

Cons:

- Continuous DC motors require gear reduction to provide the required amount of torque for most robotics applications.

- They come with poor standards in mounting arrangements and sizing.

Stepper Motor
Pros:

- Unlike DC motors, stepper motors don't require gear reduction for powering the robot at lower speeds.

- They cost less when you purchase them on surplus.

- Stepper motors have a dynamic braking effect, which means that they do not turn when their coils are energized.

Cons:

- Stepper motors cannot perform well under varying loads.

- Stepper motors are not suitable for moving robots on uneven surfaces.

- They need a special driving circuit to provide stepping rotation.

Servo Motor (Radio-Controlled)
Pros:

- The servo motors are the least expensive of those used for gear motors.

- They provide continuous rotation (with some modifications) and precise angular control.

- They are available in various sizes with standard holes for mounting.

Cons:

- They need to be modified for continuous rotation.

- Like stepper motors, servos need special driving circuits.

- Though there are servos with more power, the practical weight limit for them is around 10 pounds.

Note: Here, we have only discussed about the radio-controlled servos, though there are other types. These are the widely used servo motors and are the least expensive.

Remember that the motors can be found in various sizes and you should use the motor of the right size, depending on its workload. The points given below will help you to choose the right motor for the job.

- Small motors are intended for applications that value compactness over torque. Though there are compact high-torque motors available, they are expensive because they contain high-efficiency bearings, powerful rare-earth magnets and a few other features, adding cost.

- Large motors usually produce higher torque but they require a higher current, which in turn requires high-capacity batteries along with better and bigger circuits, which won't burn out or overheat under high load. It is very important to match the motor size with the rest of the robot. In cases where size is not a problem, do not use a large motor and overload a small robot.

- After deciding the motor size, you should compare the torque available after gear reduction. The gear reduction amount is directly proportional to the increase in torque. For instance, if the reduction ratio is 3:1, the increase in torque will roughly be about three times (or maybe slightly less due to frictional losses).

Chapter 5

Materials for Building a Robot

After deciding what type of robot you want to build, you should decide on the method of construction and the materials to be used. It may surprise some of you, but building a robot is not as simple as going to the garage and getting some pine.

You can choose the materials for building your robot from scratch. You can use materials like sheet metal or plywood. If you want to avoid all the hassle, you can go get a ready-made product for using it as your robot's base. Toys, hardware items, and inexpensive housewares can be used for building your robot in a more economical and faster way.

Build Your Robot from Scratch Using Wood, Plastic, Metal, and Composites

Building a robot from scratch means using raw materials like wood, plastic, metal, and composites as the body of the robot. Usually, the body is made by cutting the required shape from a larger piece. The common body shapes for robots are square, round, and oval.

Note: Note that there is a difference between building a robot from scratch and building a robot using ready-made commercial materials. Remember that when you are building a robot from scratch, you will require a minimum set of tools, including screwdriver, drill, and a saw. Apart from these, you will obviously need a place for building your robot. Scratch building is for people having these tools and a place to ply their craft. People with a limited number of tools or people without a workshop should consider using toys or adapted parts for building their robots. You should match the available resources with the robot construction method.

Choosing the Material

You should choose the raw construction materials based on their suitability for your job and their shaping and matching requirements, but not on their availability and price. Consider the skills and tools you have and get the materials accordingly. Using cheap materials for building the body of your robot is not advised. They might be freely available but they are not reliable. If you care about your temperament and time, avoid using materials that cause frustration and work. It is wise to keep your work as simple as possible.

Wood

Of all the materials used for building a robot, wood is the least expensive. Wood is also easy to work with ordinary tools. You might need some skills for working with wood but you won't need special tools. Soft woods like fir and pine are bulky compared to their weight; using them is not advised. Using hardwoods like birch or ash is a really good option as they are a strong and compact. Using oak is not a very good option because its density makes it hard to drill and cut.

Hardwood plywood is designed for building models because it won't delaminate and is sturdy. The only disadvantage is its cost. Plywood can take impacts to an extent as they are resistant to flaking and cracking. You can also use exotic woods that are meant for millwork. These make your robot look good and unique. They cannot be used with robots meant for rugged use, not because they are not strong, but because you will be spoiling their look.

Hardwood doesn't mean that the wood is actually hard. It is just a term given for woods obtained from deciduous trees. Similarly, softwoods are obtained from non-deciduous trees, which are basically evergreen trees. Balsa and cottonwood are both hardwoods as they are obtained from deciduous trees. They are strong but they have low density; this means that they are soft enough to work with. Balsa is both lightweight and strong, making it ideal for building model airplanes. It can be used for building robots and for strengthening the structure.

Plastic

Many manufacturers use plastic because of its molding capability. Plastic parts are made by a manufacturing technique called injection molding. Amateur robot builders cannot use this process for making robot parts. Instead, plastic can be bought from sign makers, specialty plastic retailers, and from home improvement stores in the form of rods, bars, sheets, etc.

There are thousands of plastics specifically designed for particular applications. For building a robot, plastics can be used and they are available readily. Unlike metals and composites, plastics are affordable. They can be easily modified using standard tools. Several different types of plastics are used for building robots:

- **Acrylic**: These are primarily used for functional or decorative applications like salad bowls or picture frames because of their durability. For this reason they can also be used for making robot bodies. Before using them you should consider their limitations for impact shock and weight.

- **Polycarbonate**: These are similar to acrylics in looks but are a lot stronger. Their density is greater than acrylics, which makes them hard to work with.

- **PVC**: PVC plastics are readily available in various shapes, sizes, colors, and thicknesses. Since they are manufactured using a gas expansion process, they are very light. You can cut out the desired shape or structure from these rigid sheets. These can be drilled, cut, and even sanded, just like wood. PVC can be substituted for wood.

- **Urethane Resin**: These are common components used in "casting" plastics like fiberglass. These can be found in readily available shapes and in already cured rods, bars, and other shapes. You can mold your own shapes by using liquid resin.

- **Acetal Resin**: This is commonly referred to as engineering plastic. It is usually used in applications that do not require the strength of metals like aluminum or steel for milling and turning parts. These plastics are softer than metals, which makes them easy to work with. These are surprisingly strong and dense. You can use a lathe for making Acetal resin components for building a robot. You need not worry if you don't have access to a lathe machine. These can be cut with some effort. They are readily available in the shape of sheets, bars, and rods.

For constructing the base of your robot, rigid PVC is a good choice. All of these plastics can be bought from home improvement stores. You can buy Acetal resin and PVC sheets from specialty retailers.

Metal

Metal is another type of material used for building a robot. Metal is expensive compared to other materials used for robots. It weighs more compared to other materials and it is hard to work with. You should have the proper skills and tools for dealing with metal. Metal can be used in situations where strength is chosen over mobility. For moving a metal body, the robot will need a powerful motor and a bigger power source for powering the motor. It is ideal for robots designed for rugged outdoor use. You must use metal if you are designing your robot for robot-bashing, where your robot will be fighting other robots. It can also be used in situations where your robot needs a strong but compact body.

Steel and aluminum are the most commonly used metals for building robots. It is easy to work with aluminum because it is several times softer than steel. The bodies of the robots can be made with rod, bar, sheet, and other shapes. In general, there are two approaches for constructing robots with metals. They are frame base and shape base.

- **Frame Base**: A frame can be used as the robot's base to give it support. The frame can either be box shaped or flat. Flat frames make a convenient platform for placing the robot components such as motors, batteries, etc., because of its sides and corners. The box shaped frame is a 3-D box having six faces. The box frame is ideal for building large robots or robots that require extra strength for supporting their heavy components.

- **Shaped Frame**: The shaped frame this is nothing but a piece of metal, cut in the robot's shape. The metal must be strong enough to support the weight of batteries without flexing or bending. For smaller robots, 1/32" aluminum or steel of 22 or 24 gauge usually does the job. When thicker sheets are used, there will be a dramatic increase in the weight of the robot. You can actually build robots just by using a metal sheet on wheels capable of holding heavy weights.

Composites

There are three primary forms of composites:

- Any laminated material in the form of a sheet, combining paper, wood, plastic, or metal, for using the intrinsic properties of each, with the intention of increasing the strength or rigidity or both. One of the most commonly used laminate composites is made from two pieces of paper that have springy foam between them. There are other laminates made by combining metal and wood, paper and plastic, or any other combination.

- Materials made of resin and fiberglass. In some composites, carbon, fabric, or metal will be added to provide extra strength to the resin.

- Materials using graphite or carbon for strength. These materials sometimes have other components.

> Carbon composite tent poles are a good example of these kinds of composites. They are incredibly strong, flexible, and lightweight.

Why should we use composites for making robots? The reason is their weight-to-strength ratios. Not all components have good weight-to-strength ratios. Foam board composites are lightweight but they are not very strong. They can be used for reinforcing plastic or wood parts, or for creating mockups. You can work with them just with a straight edge and knife. Other composites like carbon composites are both strong and lightweight. These components require special tools for working on them.

The availability and cost of the stronger composites are the disadvantages. Most composites can only be obtained from industrial suppliers and specialty retailers.

Chapter 6

Tips

Follow the below tips for building your robot.

1. **Decide Your Purpose:** You need to find a starting point with a specific goal in mind. You should know where to start and where your work ends. It will be a lot easier if you have a clear picture of the things to do. You can make your things move by focusing on something specific. This picture will also keep you motivated by reminding you of your idea. After making the picture of your robot, you should decide on its features. You will have to create a list of all the specifications for your robot. It might also take a few

prototypes. Based on the specifications, you will need certain skills, budget, and tools.

2. **Determine Your Strategy:** It is better to start with Internet research or with the basic exercises. Start learning about simple robotic projects like obstacle-avoiding robot, line-following robot, etc., as they give you a rough idea with your work.

3. **Keep Your First Try Simple:** If you search online, you will find many robotics projects with instructions on how to do them. Most of the articles are explained step-by-step in detail. They are very easy to follow and you can actually copy these projects for trying your first one. Just make sure that they are not copyrighted.

4. **Ask Questions, a Lot of Them:** You can actually put some questions to your colleagues, family, friends, and your professors seeking help about your plans and project. You can clarify your questions on programming, mechanics, or electronics. Their responses can affect the final outcome of your project if you take them seriously. Keep asking yourself why

you need to do something. Why should you read anything, why should you not learn Java but C++, why should you use kits for building a robot, and so on. And don't just stop with questions; try to find the right answers.

5. **Learn to use Different Tools:** To build a robot, you might have to work with a number of tools, including classical tools like a hammer, screwdriver, etc., electronic tools like a drill, hot air guns, etc., and digital tools like an electric soldering gun, digital multimeter, slotted screwdriver, wire stripper, etc.

6. **Learn from Courses:** You can take free online tutorials to get a basic idea on the reduction of engineering robots. Many famous robotics experts publish their works free of cost on the Internet. You can go through these to get a rough idea.

7. **Keep the Weight at Minimum:** Always try to keep the weight of the robot at the minimum. You can do this by using lightweight parts. Use metal only if needed. Maintaining a good power-to-weight ratio will save you a lot of battery.

8. **Get the Parts:** You will need a long list of parts for building a robot. These parts include wires, tracks or wheels, frames, motors, sensors, etc. Prepare a list of all the parts beforehand and purchase them at once.

9. **Documentation:** It is a very good idea to document the work you do. You might sometimes need to redesign your work and proper documentation will help you a lot in such cases. It is a good idea to document every line or every block of code in comments. Comments make your work easier to read, for you and also for others.

10. **Try Solving the Problems Yourself:** It is only normal to encounter problems while designing something. And not all of these problems can be solved easily. Some problems will obviously be more complex and severe than others and, to solve these you might need the help of others. Before asking anyone for help, try to solve the problem by yourself. You might actually learn a lot of things in this process by reading and searching about the encountered problem.

11. **Safety First:** One of the common sources of injury at work places is manual handling. There are a few risks involved when building robots. You might have to use power tools like an angle grinder, heat gun, power drills, etc., and you should take extra care while working with them. You cannot recklessly use batteries because they might explode if not treated well. Among the most commonly used batteries for robots are Li-Po batteries, which come with the risk of exploding under some conditions. You should take appropriate protective measures like looking for potential hazards by walking through your workplace. By using protection, you can eliminate or reduce the risks.

12. **Budget:** Building robots is not always cheap and, if you wish to build a specific type of robot, you might have to invest something in software, tools, kits, or parts. The more components you add, the higher the cost goes. Most of the robots need sensors, such as infrared sensors, bump sensors, light sensors, etc., and they will cost you some bucks. You need to do a bit of research to save some money. Usually, online

stores are cheaper than local stores. But if you find a local dealer giving the same part for a lesser price, go for it. If possible, ask him to test the part. You can build robots like optical-avoiding robots, line-following robots, etc., for less than one hundred dollars. The cost of your robot increases with its complexity.

13. **Try and Retry:** For anyone, the first project might be the hardest and for gaining experience takes some effort and time. After completing your first project, you can see that the things start moving. The effort needed will also decrease in time. There are many students, hobbyists, and kids who were excited to build their first robot, who probably failed at some point. This is where you shouldn't get disappointed. Put your time and effort into it again and try. The payoff will eventually come to you.

14. **Share Your Experiences:** The Internet changed the way we share knowledge. You can learn almost anything from the Internet and you are simply a few clicks away. You can take help and help others. If you have information on something, you can share it

share in forums or in specific community groups in the form of articles or comments. This might help other robotics enthusiasts. The value of your experience increases if others find it useful. The most important places for sharing are where you document or explore. If you did something or if you find something interesting, do not forget to post it for others.

15. **Be a Perfectionist:** You will have to face poor organization and planning, along with the results. This is where things might become complicated. You should try to make things perfect. Give it some time and don't plan on finishing the work quickly. It might take you some more time but a good result is what you need.

16. **Learn Electronics**: Though it is not the most fun part to learn in robotics, having knowledge of electronics is essential. It is not easy to learn robotics without having any knowledge of electronics. You don't need to have a degree in electronics but you should know the basics. There are many books online that will help you learn basic electronics.

17. **Buy Books**: Get some books to have a good start with robotics. By reading the right books, you will get to know all the important topics and concepts of robotics. It is also a good practice to follow magazines on robotics.

18. **Start Small**: Don't try to start a big project in the beginning. Don't try to begin with complex projects but instead start small ones. There are basically two types of robot builders, the people starting from scratch and the ones using kits. Know the type of person you are and start accordingly.

Applications of Robotics

In the present day, robots are used in almost every field, performing various jobs. And the dependency on robots is increasing progressively. Robots can be categorized into various types based on their areas of use.

1. **Industrial Robots**: These robots are used in industrialized manufacturing environments. They are typically articulated arms designed for particular applications such as welding, painting, material handling, packing, etc. If we sort robots according to

their applications, these robots also include automatically guided automobiles.

2. **Domestic or Household Robots**: These are the robots used in the home and they consist of a number of different gears. Examples are robotic sewer cleaners, robotic vacuum cleaners, robotic sweepers, robotic pool cleaners, and robots capable of performing various other domestic tasks. The telepresence and scrutiny of robots also come under this category if they are used for domestic environments.

3. **Medical Robots**: These are the robots used in medical institutes and medicine. The surgical treatment robots are the first and foremost. Using these in the area of medicine has brought a revolution. The robotic-directed automobiles used in hospitals and for physically handicapped people also come under this category.

4. **Service Robots**: Service robots include those that cannot be categorized into other types, by practice. Included are various robots created for exhibiting

technologies, data-collecting robots, robots used in research, etc.

5. **Military Robots:** Robots that are used for Armed Forces and military purposes fall into this category. Military robots include various shipping robots, bomb defusing robots, exploration drones, etc. Usually, these robots can also be used in exploration, law enforcement, and any other fields associated with them.

6. **Entertaining Robots**: This category is an extremely wide one. Humanoid entertainment robots, substitute pets, robosapiens, virtual woman, etc., come under this category.

7. **Space Robots:** These are the robots used for space programs. They are designed for exploration, for aiding communications, data collection, aiding astronauts, etc. Robots such as Mars explorers come under this category.

8. **Competition and Hobby Robots**: These are the robots created by hobbyists and students. They are designed for fun and learning purposes. Another

category is the competition robots where they are specifically designed for robotic competitions. Battle bots are included in this.

Peter McKinnon

Chapter 7

Circuit Schematics

One of the crucial parts of building a robot is knowing the exact plan or schematic of circuits and components that you will be using. You also need to carefully map out the design of your robot and where exactly things need to be placed in the robot for an effective utilization of the component. Here is a list of things you might want to consider before putting together a robot.

- It must be cost-efficient. There are many kits out there in the market that are expensive, but the same result can be achieved for a much lower price.

- It must not be complicated to maneuver and should be easy to put together without the use of any special equipment.

- The robot must be compatible with the programming languages that you are familiar with, so it should have a complex IDE or a programmer.

- It must also be powerful enough for expandability.

- It must not require a complicated power source.

- It should be able to navigate its way through a passage and must avoid obstacles.

Now let us see how we can meet these requirements below.

Choosing the Components

The first step you need to take is to figure out what components are required in the building of a robot. This acts as a skeletal plan for your robot. A generic robot needs a few things to be useful: a way to observe the environment, record or calculate the readings, and then be able to manipulate the stimulus. In order to meet our budget, let us construct a robot with only two wheels; to operate it, we will require two separate motors that can be operated independently. We might also need a ball caster that the robot uses to lean on to glide along the path. One of the disadvantages of these types of robot is that it will not be able to move on surfaces other than smooth floors. It will not be able to navigate in the yard or climb up or down the stairs. You might also need to learn about the various microcontrollers that can be used to program the motors and the sensors that you may add to the robot. Try to familiarize yourself with the programmer and the guide of the microcontroller development tools that will come in handy while programming or customizing the working of your robot. You need to install sensors that will help the robot detect walls, bumps, and other obstacles. And, lastly, you need to design a low-cost PCB board that will help you conglomerate the components on one single chip or a simple low-cost board that resembles an Eagle CAD.

Mechanics: Motors, Gears, and Wheels

There are various websites that offer a variety of hobby motors and robot parts but the major downside to finding the right fit for your robot is that it can get a little bit on the expensive side. It is recommended that you do extensive research before you place an order. There are plenty of sites like Pololu, Tamiya, and so on, that not only offer great deals, but also have almost any component that you might need for your robot. The 70168 Double Gearbox Kit contains a great range of gears, motors, shafts, and other mechanics that go into the building structure of your robot. These motors usually operate at 3V input, but have the ability to run higher at the expense of reduced battery and operational life. You can also fine-tune the gears in the robot by making adjustments to the gear ratios.

Microcontrollers

This is one of the most crucial parts of building your robot, since there are a fair number of microcontroller platforms available in the market. One of the obvious choices is the Arduino board, due to its compact size and agility. Some of the other controllers that you can use are the Teensy, LaunchPad, Raspberry Pi, or even the basic ATMEG Microcontroller 8051 series. Each of these has its own upsides and downsides and you need to choose the one that will be comfortable for you and also gives an efficient output. While it is possible to build the robot using Launchpad and Raspberry Pi, their size and the power requirements are a deterrent to the effective functioning of the robot. It is recommended that you go with one of the smaller versions of Arduino Mini or Teensy, which happens to be quite effective. The latest version of Teensy has a Cortex M4, which provides the robot with the necessary power. And there is an added bonus in using the Teensy board, since it comes with an onboard 500 mA regulator that is perfect to drive any sensors attached on the board.

Interaction: Sensors

While implementing a robot that has to follow a line or move around, it is important to note a variety of sensors are used in order to prevent it from dashing itself into obstacles. The line followers usually use reflectometers, which pick up the amount of light that is deflected from the ground and the voltage is varied. Once the voltage goes below the threshold limit, the robot needs to stop. The LED, photodiodes, and other light detectors are used in this type of sensor. In order to detect walls, other bumps and elevations, obstacle sensors are used, which are usually based on distance sensing. These types of sensors are available in the same website as the motors and in the form of a convenient DIP breakout form. This helps you cover the cost of shipping and, more important, these pieces are easier to solder. For the line follower, you can use the one with three sensors that helps the robot to line in the center at all times. To detect the distance, you can use a high-brightness IR sensor, since that is the most efficient when we use a lower voltage power driver circuit.

Power: Motor Driver, Battery

Every motor driver should provide a minimum of 3V to start a motor. It also needs to be scalable in case you need to upgrade the motors in the future. You can find motors of such specifications on the websites mentioned above. They have the ability to operate 0 to 11 V and at the same time supply plenty of current to the motor that you may add in the future. The typical input Teensy approves is up to 5.5 V, which means that a lithium ion battery can be used. Lithium batteries are also rechargeable and you might need a charger and a cable to support it. If you don't want to use lithium batteries you can always use two normal AA batteries that offer great power over a period of time. The disadvantages of this battery are that it is quite big and it only supplies 3V. The Teensy's linear regulator has a threshold of about 3.3V and the 3V voltage will not be able to power the device. The robot, on the other hand will still operate, due to the components chosen and Teensy can operate on lower voltage. However, the regulator on the Teensy board will still function without regulating the voltage.

Optional Items

These are the peripherals that you may want to add to your robot to customize it according to you needs. The most popular addition is the BLE device that can be included in the schematic so that the device can be operated remotely using a smart phone at any point. You can also enable the robot to follow lines and walls. In order for you to remove and add items to your robot, it is recommended that you use female headers to connect the ports on the board.

Schematic

You should draw a rough diagram of how you want your robot to look, with the addition of the above-mentioned components and where they should go. One of the main things to start with is the power supply to drive the device. You can draw up block diagrams and flow charts in order for you to organize your thoughts and your vision for the robot and give it a frame.

A schematic can also be a circuit directory of all the components used in a particular module and how they are used in the robot. The power supply, the sensors, the mechanics can all exist as different modules before you attach them all together in your robot. These diagrams are very helpful in understanding how to connect the separate entities and also helps you sort out a mistake easily.

Here are a few keynotes that you can use while creating your schematic circuitry and implementing it on a PCB board:

- Remember to add jumpers between the battery and the rest of the circuitry since this will help when you have to disconnect the power supply and you need not remove the batteries at that time. This is also useful for measuring current and protecting the circuit from reverse polarity with the help of a diode.
- The interfaces are all digital except two. There is a UART (universal asynchronous receiver transmitter) connection between the Teensy through pins 9/1 and

the nRF51. In order to control the speed and direction of the motor, the controller requires PWM (pulse width modulation), which is supplied through pins 6 and 4 of the Teensy board.

- Unless there is another use for an LED light setup, you can avoid the use of external LEDs. The LED, which is available on the Teensy, can be used for indication or even debugging.

- Avoid placing external switches or buttons, since they will only complicate the circuit. Place them only if they are an absolute necessity and will also help you lower your costs.

- While programming in Teensy through a USB port, you must ensure that there is either a small trace cut for connecting to Vin or the Vusb or be sure that the batteries are disconnected when the USB is plugged in.

Peter McKinnon

Chapter 8

How to Build a Simple Robot for Beginners

This tutorial has been designed with the complete beginner in mind. It is simple to follow and great fun to build. I have broken the tutorial down in many steps so that you can easily follow the instructions, one step at a time. This robot really isn't anything more than a wooden platform with a motor, controller, and sensor. Provided you have all the parts in hand, you should be able to knock this one out in about three hours.

Materials

- Microcontroller – whichever one you want

- Sharp GP2D12 that has a wired JST connector
- Aluminum wheels or tires
- Hitec H311 servos – 2
- Super glue
- 5 x 1 inch double-sided tape
- ¼-inch plywood
- Scrap wood
- .20-.40 tailpiece assembly for radio-controlled aircraft
- A heap of spacers, standoffs, screws, and nuts
- 9V battery
- 1/16 x 3 inch heat shrink tubing
- Standard female crimp pins – 3 (no panic if you cannot get these)

Tools

- Small screwdrivers, miscellaneous sizes

- Needle nose pliers
- Good drill and assorted bits
- Soldering iron and solder
- Saw
- Sandpaper
- Sharp pencil
- Ruler

This whole project should take between 2 ½ to 3 hours to complete and will cost around $100.

Step 1 - The first job is to modify the servos. We do this with the potentiometer method. You need to tear down the servos, locate the 90 position, and super-glue the top and bottom of the pot. Put it all back together and modify the output gear, filing off the pot shaft flush level with the case molding. Or you can use one of these methods. If you use the Hitec servo, the plastic shaft is easy to file but if you use one with a metal pot, you will need to make a change to the underside of the output gear.

Potentiometer – pot – variable resistor

Step 2 – Now it's time to cut the wood. Take your piece of ¼-inch plywood and cut it into a square of 3.25 inches – this will be the base of your robot. Next, take and a piece of your scrap wood that is 2 5/16 inches long, ½ inch thick, and ¾ inch tall. This will be your GP2D12 mount. Sand everything down gently to clean it up.

Step 3 – Lay your microcontroller on the plywood, toward the front, as a guide. Mark where the mounting holes are. Using the tail wheel mount as another guide, lay that toward the real of the wood and mark out the mounting holes. Using a 1/8 drill bit, drill all of the holes.

Step 4 – Super-glue the servo mount and the microcontroller mounts to the base of the plywood. Make sure they are centered and flush with the front edge of the wood.

Step 5 – Put your servos into their positions behind the mount, and you can see that the wires hit the backside of the mount. To let them pass through, you need to drill some holes. Make sure that the servo is positioned so that the output shafts are toward the front of the wood.

Use the servo to mark out where the holes for the wires should be and then, using the pencil and ruler, work out where those holes need to be on the front of the mount. Use a 5-1/6 drill bit to drill the holes but be careful that you do not split the wood.

Alternatively, you can drill a small hole at the bottom of the servo, during the modification process, and the wires can be passed through there instead.

Step 6 – Use 4-40 x ½ inch screws and nuts to fasten the tail wheel assembly to the middle rear of the wood. Make sure the screw heads are countersunk so they don't get in the way of the servo when you install it.

Step 7 - Mount your standoff now, before you put the servo in. To do this, use a 4-40 x 1 inch screw, a 4-40 nut, and a 4-40 x ¼ inch standoff for each of the microcontroller holes. Ensure that the screw heads are countersunk.

Step 8 – To install the servos, super-glue them to the base of the plywood, making sure they are against the servo mount. Push the wires through the holes in the mount.

Step 9 – Fix the microcontroller into the standoffs. You might need to use a washer on each of them to bring the microcontroller above the tail wheel assembly.

Step 10 – Now you can install the wheels you have chosen and make any adjustments to the tail wheel as needed.

Step 11 - Use double-sided tape to attach the battery in the space that should be between the servos and the tail wheel, on the bottom of the plywood. The battery needs to be installed on edge, otherwise it will not fit

Step 12 - Now it is time to plug in the servos – the left one plugs into port 31 on the right 3-pin port and the right one goes to port 30.

Step 13 – Fix the GP2D12, using super glue, over the wires and holes that are on the front mount. The wires should be up and there should be a little bit of give in the plastic casing so that the wires can slot in behind.

Step 14 – This is the hardest part of the whole project. Put a piece of heat shrink tubing onto each wire on the GP2D12 and slide it up. Use a single crimp pin receptacle on each one, soldering it in place. If you are very adept at soldering, you could just solder the wire from the end right to the tip of the microcontroller pins and then put the heat shrink tubing on top for protection.

Otherwise, slide the tubing so that it is flush with the crimp pin and use a little heat, with either a lighter or a heat gun, to make the tube shrink. Look for a 5-volt pin that is open on the microcontroller and plug in the red GP2D12 wire. Now locate any GND pin that is open and plug in the black GP2D12 wire. Finally, locate pin 3 and plug in the white GP2D12 wire.

Step 15 – Use the directions that came with your microcontroller and, when you get to the editor, add the following code. It must then be compiled and sent to your robot.

```
Dim     Servo_Right    As    New    oServo
Dim     Servo_Left     As    New    oServo
Dim     SRF04Servo     As    New    oServo
Dim     GP2D12         As    oIRRange(3,8,cvOn)
Dim     SRF04          As    oSonarDV(8,9,cvOn)
```

SRF04Servo.IOLine=29 'Set the servo to use I/O Line 30.

SRF04Servo.Center=28 'Set the servos center to 28. (See manual)
SRF04Servo.Operate=cvTrue 'Last thing to do, Turn the Servo on.
Servo_Right.IOLine = 30 'Set the servo to use I/O Line 30.
Servo_Right.Center = 28 'Set the servos center to 28. (see manual)
Servo_Right.Operate =cvTrue 'Last thing to do, Turn the Servo on.
Servo_Left.IOLine = 31 'Set the servo to use I/O Line 31.
Servo_Left.Center = 28 'Set the servos center to 28. (see manual)
Servo_Left.Operate =cvTrue 'Last thing to do, Turn the Servo on.
'--
' End Create and Setup Objects
' Main routine is your primary routine called upon power up!
'--
Sub main()
Do
Call IR
Call Forward_All

```
Call                                    SServo
Loop
End                                      Sub
'-----------------------------------------------------------
'       End      of      Main       routine
'    Start   of    Drive   System   Subroutines
'-----------------------------------------------------------
Sub                                Spin_Left()
Servo_Left.Invert=0
Servo_Left              =              62
Servo_Right             =              60
End                                      Sub

Sub                                Spin_Right()
Servo_Right.Invert=1
Servo_Left              =              62
Servo_Right             =              60
End                                      Sub

Sub                                Forward_All()
Servo_Right.Invert=0
Servo_Left.Invert=1
Servo_Right             =              60
```

```
Servo_Left                    =              62
End                                          Sub


Sub                                      REVERSE()
Servo_Right.Invert=1
Servo_Left.Invert=0
Servo_Right                   =              60
Servo_Left                    =              62
End                                          Sub


Sub                                        STOP()
Servo_Left                    =               0
Servo_Right                   =               0
Call                                        SServo
End                                          Sub
'-----------------------------------------------------------------
'    End    of    Drive    System    Subroutines
'        Start        of    Sensor    Subroutines
'-----------------------------------------------------------------
Sub                                        SServo()
SRF04Servo.Position=15
SRF04.Operate.Pulse(1,1,250)
If              SRF04.Value<128              Then
```

Call Spin_Right
Else Call Forward_All
End If
ooPIC.Delay=600

SRF04Servo.Position=31
SRF04.Operate.Pulse(1,1,250)
If SRF04.Value<64 Then
Call REVERSE
Else Call Forward_All
End If
ooPIC.Delay=600

SRF04Servo.Position=46
SRF04.Operate.Pulse(1,1,250)
If SRF04.Value<128 Then
Call Spin_Left
Else Call Forward_All
End If
ooPIC.Delay=600

SRF04Servo.Position=31
SRF04.Operate.Pulse(1,1,250)

```
If          SRF04.Value<64           Then
Call                            REVERSE
Else            Call         Forward_All
End                                   If
ooPIC.Delay=600
End                                  Sub

IR()
If          GP2D12.Value<64          Then
Call                               STOP
End                                   If
End                                  Sub
'---------------------------------------------------------------
'       End      of     Sensor    Subroutines
'---------------------------------------------------------------
```

Chapter 9

A Simple Touchless Sensor Robot

This project is quite simple and you should end up with a very simple robot that you control using a keyboard or with a Leap Motion sensor. The robot runs on a microcontroller programmed with a version of BASIC 1 and the best chip to use is a PICAXE 20M2. This is because it is relatively cheap to buy and does not struggle with picking up the coding language. However, as it is not already mounted on a board, we will need to do a little bit of fiddling around with a breadboard.

The robot is controlled with Python and PySerial is not supported on the Mac, so you will need to use a Windows

PC. You can still build the robot but you will need to find another way of controlling it.

Hardware Requirements

- PICAXE 20M2 microcontroller – one piece
- Serial programming cable – one
- 3.5 mm socket – one
- 10K resistor – one
- 22K resistor – one
- 480 Ohm resistor – one
- LED – one
- 5V IC power resistor – one
- Capacitors – 100uF (two pieces) and 0.1uF (two pieces)
- Power switch – one
- 9V battery including wires to attach into a breadboard
- Wire

- Breadboard – one piece

- Computer to enable you to program the PICAXE

In order for the PICAXE to be able to control your robot, you will need:

- Dual H bridge motor driver IC – L293D – one

- DC motors with gearboxes – two pieces

- Wheels to attach to the motors – two pieces

- Alligator jumper wires – four pieces

- Robot body – a piece of Plexiglas or similar, 4 ¾ inch by 4 ¾ inch

- More wire

- Motor power battery that has wires to attach to the breadboard. Should be between 4.5V and 36V as this be the source of the power for the motor

- Scotch tape

Software Requirements

As you would expect, the PICAXE needs a little help to get it running, and that means doing a little programming. Don't worry; I will give you the code that you need. The PICAXE has a decent programming environment, which makes it relatively simple to program the chip. To do that you are going to need:

- The latest version of PICAXE Editor for Windows
- AXEpad for Windows, Mac, or Linux
- AXE027 drivers for the USB cable – Windows and Mac only, not required on Linux

To be able to program the controllers, you will need:

- Python 2.7 – latest version
- PySerial for Python 2.7 – latest version – Not supported for Mac
- PyGame for Python 2.7 – latest version

To control the robot using a Leap Motion sensor, you will need:

- Leap Motion SDK, which you can get from the developer. It does require you to sign up for a free developer account

Step 1 – To program your PICAXE chip, it has to be connected to a computer and this is where your serial cable comes in. This cable connects to the PICAXE using a 3.5 mm jack, the same as many audio devices. ***DO NOT USE THIS CABLE FOR ANYTHING OTHER THAN PROGRAMMING OR COMMUNICATION WITH THE PICAXE – IT IS NOT INTENDED TO BE CONNECTED TO ANY AUDIO DEVICE.***

The socket that this cable plugs into can be a little difficult to keep in the breadboard, so you should solder wire onto the socket terminals. This will stabilize the connection to the PICAXE and it will also put the socket in a better position. The left pins are for incoming data, the right pins are for GND, and the top is for outgoing data. Color-code your wires if you want.

When you are soldering the wire on, ensure that all the pins grouped together are soldered onto the wire.

Step 2 – The next step is to add the 5V regulated power onto the breadboard and then build the PICAXE circuit. To set up the power supply:

- At the top of the breadboard, put the LM7805 power regulator, making sure that the aluminum piece faces to the right. The wires need to be added between row 3 on the left and the left VCC bus, between row 2 right and row 2 left, between row 4 right and row 2 right and between the right GND bus and row 4 right.

- Put the power switch in between row 1 right and row 1 left. If there are any extra pins, ignore them.

- Connect together the VCC buses and connect the GND buses together, to give you GND and power along the breadboard.

- Put a 100uF capacitor between row 2 left and row 1 left. Make sure that the negative side faces row 2 left. Put another 100uF capacitor between the left GND bus and the left VCC bus, making sure that the negative faces the left GND bus.

- Repeat the above step with the 0.1uF capacitors.

Adding the PICAXE:

- Put the 10K resistor between row 4 left and the left GND bus.

- Put the 22K resistor between row 8 left and row 4 left.

- Make sure the PICAXE is situated so that pin 1 on the top left is on row 7 left.

- Put wires between row 7 and GND right and row 7 and VCC left.

- Put the green data wire from the download socket on row 4 left, the yellow wire onto row 8 right, and the blue GND wire on the GND.

- Put the 480-ohm resistor between row 17 left and row 11 left.

- Put the LED between the GND and row 17 left.

- Put the 9V battery with the positive on row 1 right and the negative on row 2 right.

You should now have a PICAXE circuit. Just have a quick check over all the wiring and make sure it is all where it should be.

Step 3 – Now that your PICAXE chip has been wired, the next step is to program it. Open up the PICAXE website and download the right drivers for the serial cable you are using. Once you have done that, you need to install the drivers.

If you are using a Linux machine or a Chromebook, these drivers are already installed by default. The driver is designed to tell the Mac or Windows operating systems how they should use the USB cable. This is just a one-off install process that must be done the first time you use the cable. Once you have installed the driver, your computer will remember the settings the next time the cable is used. Installing is quite easy:

- Make sure that your serial cable is NOT plugged into your computer

- Download the correct drivers

- Run the installation program

- Plug in the cable and then follow all the onscreen instructions

A few tips for proper installation:

- Put the cable into your computer a couple of seconds before you start the PICAXE software.

- However many times you plug the serial cable in to your computer, always use the same USB socket. That way the virtual COM port setting will always be the same.

When you plug the cable in you will be given a serial port ID; write this down, as it will be required to ensure that the PICAXE editor recognizes the cable and for when you configure the control program for your robot.

Step 4 – Next, you need to download the editor and install it. You need to do this with PICAXE editor for windows and with AXEpad. As soon as you have your editor, you will be ready for programming. The code I have given assumes that you are using Windows 7 and PICAXE Editor 6. If you are using something different, the code will work but the environment is going to look a little different.

Step 5 – Now we have the hardware set up and the software but, at this stage, we do not know if our PICAXE works. The best way to find out is to write something called a "blink" program, which will switch our LED on and off. You can either write your own "blink" program or you can use this one. Input it into your editor, save it, and then open it with the PICAXE editor.

main:

    high C.5

    pause 1000

    low C.5

    pause 1000

    go to main

Plug in your serial cable.

If you are using Editor 6, locate "COM Port" on the left hand side of the GUI. Click on it and then click on "Refresh COM Port." If the ID of your serial cable appears, you are nearly ready to program your PICAXE. Plug in the 3.5 mm jack on the cable to the socket on the breadboard and then click on "PICAXE Type." Change the name to that of your specific chip.

The last step is to click on the button that says "Program" located at the top of the interface. Follow that by clicking PICAXE power switch to turn it on. If all goes as it should, PICAXE should begin downloading the program you downloaded and the LED should begin to blink. If it does, well done, you have successfully programmed your PICAXE chip.

Step 7 – Now you know your PICAXE is working as it should be, you need to add in the motor controller chip. Here is how to do that:

- Put pin 1 of the L293D H bridge on row 18 left.

- Put wires between the VCC and row 18 left, the VCC and row 18 right, and the VCC and row 25 right.

- Put wires between row 21 right and 22 right, and in between row 21 left and 22 left.

- Put A wire between the GND and row 22 left, and the GND and row 22 right.

- Put wires between row 21 left and 26 left.

- Put wire between row 14 left and 24 left, and in between row left and 19 left.

- Put wires between row 12 left and row 24 right, and in between row 1 left and row 21 right.

- Put the motor power battery IN, making sure the positive goes to row 25 left and the negative goes to row 26 left.

At this stage, the PICAXE can control the L293D, so give the wiring a check over and make sure it is all OK and in the right places.

Step 8 - Right now, we have a working controller but we do not have any robot to control with it. Get the Plexiglas you are using as your robot body and attach the motor to it, using either transparent tape or two-part epoxy. You can also drill a couple of small holes to the front of your robot body, put small bolts through the holes, and secure them in place with nuts on either side. Attach the wheels and your robot body is finished.

Step 9 – The hardware is almost finished, we just need to add the breadboard and the batteries to the body of your robot. You can use transparent tape again to attach the batteries and breadboard, or you can use something a bit stronger if you want. Do keep it in mind that, at some point, you will more than likely want to take the breadboard off so you should probably avoid using epoxy to fasten it on with.

The next step is to get four wires and put them in at these points:

- Row 20 left
- Row 23 left
- Row 20 right
- Row 23 right

Take your alligator jumper wires and clip one end of each wire to one end of each of the four wires you just attached. The other ends of the jumper wires are going to connect to your motor like this:

- Row 20 left to left motor negative

- Row 23 left to left motor positive
- Row 20 right to right motor positive
- Row 23 right to right motor negative

Tape the jumper wires down so that they cannot get in the way of the wheels and that completes the hardware build.

NOTE – If you have used different wheels, motors, and gearboxes than the ones I recommended, you might have to connect the motors a little differently. If you find that your motors are going backward, swap the way the jumper wires are connected to the motor wires, keeping in mind that the left jumper goes to the left motor and the right jumper goes to the right motor.

Step 10 – The hardware is finished so now we need to program the PICAXE. This script is somewhat more complex than the "blink" program. It will use the serial port; this is a good reason why you should use the PICAXE and not another alternative. The PICAXE suffers very little data loss, whereas the Arduino is much worse. That makes the PICAXE far more stable and reliable.

Download this file and follow the previous instructions to program the PICAXE. Once the program has been downloaded to the PICAXE, the LED should blink again, just to let you know that it works. You can now switch off your robot and put it to one side.

Step 11 - While this little robot that you have made is pretty cool, it does need some help – it cannot be controlled without it. You need a computer and a control program, which uses Python 2.7 in order to run. Make sure that you have Python 2.7 installed. To do this, open up the command terminal and type in "Python" at the command prompt (do not include the quote marks). Press enter – if Python is installed, a message will appear and your prompt will change to >>>. If Python is not installed, you will see an error message. If it isn't installed, download it from www.python.org.

When you have Python 2.7, head over to www.pygame.org. Download and install the most up-to-date version of Pygame for Python 2.7. Lastly, head over to www.wourceforge.net/projects/pyserial and download the most up-to-date version of PySerial for Python 2.7.

That is all the software that you need to be able to run your robot.

Step 12 – Open up either your Python compiler or a text editor, like Notepad (not Microsoft Word, as it is not a text editor) and input the following code. You do need to make a few changes to it so, as you go, change the following details:

- Make sure you input your own serial port number – this is the port number that was assigned to your serial cable earlier on.

- If you are running on Linux, look at the commented lines underneath the port number line – remove the # from the start of the last line in the last set and then add a # to the start of the last line in set 2.

This is the code that you need:

```
"""----------------------------------"""

"""|PYGAME AND PICAXE MICROCONTROLLER|"""

"""----------------------------------"""

#this is the port number of your serial port
```

```python
portNumber = 4

#use this line if you're running Linux:

#port = '/dev/ttyUSB' + str(portNumber - 1)

#otherwise, use this line if you're running Windows:

port = portNumber - 1

"""------IMPORTANT! YOU DO NOT NEED TO EDIT ANYTHING BELOW THIS LINE!-------"""

"""----------------------------------------------------------------------"""

#import modules

import pygame, sys, random, serial

from pygame.locals import *

pygame.init()

print("Initialized Pygame")

#create and name frame

width = 300
```

```python
height = 300

canvas = pygame.display.set_mode((width, height))

pygame.display.set_caption("Pygame Robot Controller")

#create FPS clock

fpsClock = pygame.time.Clock()

FPS = 30

#variables

serDataL = "0"

serDataR = "0"

serDataOld = "00"

running = True

#connect to serial port

ser = serial.Serial()

if not("/dev/ttyUSB" in str(port)):

    print("Connecting to serial port COM" + str(port) + "...")

else:
```

```python
    print("Connecting to serial port " + str(port) + "...")
ser.port = port
ser.baudrate = 4800
ser.open()
while running:
    for event in pygame.event.get():
        #'x' the window
        if event.type == pygame.QUIT:
            running = False
        #movement handlers
        if event.type == pygame.KEYDOWN:
            if event.key == pygame.K_RETURN or event.key == pygame.K_ESCAPE:
                running = False
            elif event.key == pygame.K_w:
                serDataL = "1"
```

```
    elif event.key == pygame.K_s:
      serDataL = "2"
    elif event.key == pygame.K_i:
      serDataR = "1"
    elif event.key == pygame.K_k:
      serDataR = "2"
    if event.key == pygame.K_SPACE:
      serDataL = "9"
      serDataR = "9"

  elif event.type == pygame.KEYUP:
    if event.key == pygame.K_w:
      serDataL = "0"
    elif event.key == pygame.K_s:
      serDataL = "0"
    elif event.key == pygame.K_i:
```

```
      serDataR = "0"
   elif event.key == pygame.K_k:
      serDataR = "0"
   if event.key == pygame.K_SPACE:
      serDataL = "0"
      serDataR = "0"
#decipher positions, then write serial data to PICAXE
serData = serDataL + serDataR
if serData != serDataOld:
   if serData == "00":
      serDC = "0"
   elif serData == "10":
      serDC = "1"
   elif serData == "20":
      serDC = "2"
   elif serData == "01":
```

```
        serDC = "3"
    elif serData == "02":
        serDC = "4"
    elif serData == "11":
        serDC = "5"
    elif serData == "21":
        serDC = "6"
    elif serData == "12":
        serDC = "7"
    elif serData == "22":
        serDC = "8"
    elif serData == "99":
        serDC = "9"
    #print data and send to PICAXE
    print serData + " " + serDC
    ser.write(serDC)
```

```
    serDataOld = serData

   #tick clock, and update and clear display

   fpsClock.tick(FPS)

   pygame.display.update()

   canvas.fill((0,0,0))

#close serial port and pygame

print("00 0")

ser.write("0")

ser.close()

print(" ")

print("Done.")

pygame.quit()

sys.exit(0)
```

Once this has been input, you can plug in the serial cable but don't connect to the robot. First of all, you need to ensure that the program isn't going to throw up any errors.

Step 13 - Provided you got no errors, you are ready to run your robot. Make sure that the robot is switched off and then make sure that the controller program is off. Plug in the serial cable to the robot and the computer and switch the robot on. Open up the control program and wait – a black frame should appear. When it does, press on the spacebar and watch.

If all went as it should have, your robot should turn left and it should then move forward. If that happened, your robot is now receiving data via the computer and you can now have a go at driving it, using the keyboard for controlling it as follows:

- Move the left motor forward with "W"
- Move the left motor backward with "S"
- Move the right motor forward with "I"
- Move the right motor backward with "K"

Entering autonomous mode is done by pressing on the spacebar.

Step 15 – Now that we now it is all working, we come to the real point of this robot – the touchless control. We need to give our robot interface a sophisticated and touchless sensor. To do this, you are going to need:

- A Leap Motion sensor
- The Leap SDK – this is free to download from www.developer.leapmotion.com/. First, you will need to sign up for a free developer account, which gives you access to a lot of cool stuff.
- The Leap Motion-compatible Python controller

Plug your sensor into a USB port on your computer – make sure that you do not use the same one as you do for the PICAXE serial cable – and wait for the right drivers to install. When you have done that, input this code into your text editor and give it its own folder:

"""---------------------------------------"""

"""|LEAP MOTION AND PICAXE MICROCONTROLLER|"""

"""---------------------------------------"""

```
#this is the port number of your serial port
portNumber = 4
#use this line if you're running Linux:
#port = '/dev/ttyUSB' + str(portNumber - 1)
#otherwise, use this line if you're running Windows:
port = portNumber - 1
"""-----IMPORTANT! YOU DO NOT NEED TO EDIT ANYTHING BELOW THIS LINE!------"""
"""--------------------------------------------------------------------"""

import Leap, sys, thread, time, serial
from Leap import CircleGesture, KeyTapGesture, ScreenTapGesture, SwipeGesture

#setup serial port interface and kill any action
ser = serial.Serial()
if not("/dev/ttyUSB" in str(port)):
  print("Connecting to serial port COM" + str(port) + "...")
```

```
else:
    print("Connecting to serial port " + str(port) + "...")
ser.port = port
ser.baudrate = 4800
ser.open()
#serial data variables
serDataL = "0"
serDataR = "0"
serDataOld = "00"
class LeapMotionListener(Leap.Listener):
    fingerNames = ["Thumb", "Index", "Middle", "Ring", "Pinky"]
    boneNames = ["Metacarpal", "Proximal", "Intermediate", "Distal"]
    stateNames = ["STATE_INVALID", "STATE_START", "STATE_UPDATE", "STATE_END"]
    def on_init(self, controller):
```

```
    print("Initialized Leap Motion libraries!")
   def on_connect(self, controller):
    print("Leap Motion sensor connected!")
controller.enable_gesture(Leap.Gesture.TYPE_CIRCLE);
controller.enable_gesture(Leap.Gesture.TYPE_KEY_TAP);
controller.enable_gesture(Leap.Gesture.TYPE_SCREEN_TAP);
    controller.enable_gesture(Leap.Gesture.TYPE_SWIPE);
   def on_disconnect(self, controller):
    print("Leap Motion sensor disconnected.")
   def on_exit(self, controller):
    #stop PICAXE activity, then close serial port
    if serDataOld != "oo":
      print "oo o"
    ser.write("o")
    ser.close()
```

```
    print("Done.")
  def on_frame(self, controller):
    frame = controller.frame()
    if not len(frame.hands) < 1:
      for hand in frame.hands:
        global serDataL, serDataR, serDataOld
        #is hand left or right?
        handType = "Left Hand" if hand.is_left else "Right Hand"
        #define normal, direction, pitch, roll, and yaw
        normal = hand.palm_normal
        direction = hand.direction
        pitch = (direction.pitch + Leap.RAD_TO_DEG)
        roll = (direction.roll + Leap.RAD_TO_DEG)
        yaw = (direction.yaw + Leap.RAD_TO_DEG)
        if hand.is_left:
```

```python
        #middle position
        if pitch <= 57.7 and pitch >= 57.2:
            serDataL = "0"
        #forward position
        elif pitch <= 57.2:
            serDataL = "1"
        #back position
        elif pitch >= 57.7:
            serDataL = "2"
    else:
        #middle position
        if pitch <= 57.7 and pitch >= 57.2:
            serDataR = "0"
        #forward position
        elif pitch <= 57.2:
            serDataR = "1"
```

```
    #back position

    elif pitch >= 57.7:

        serDataR = "2"

    #decipher positions, then send one char serial data to
PICAXE

    serData = serDataL + serDataR

    if serData != serDataOld:

     if serData == "00":

        serDC = "0"


      elif serData == "10":

        serDC = "1"


      elif serData == "20":

        serDC = "2"
```

```
elif serData == "01":
    serDC = "3"
elif serData == "02":
    serDC = "4"


elif serData == "11":
    serDC = "5"


elif serData == "21":
    serDC = "6"


elif serData == "12":
    serDC = "7"


elif serData == "22":
    serDC = "8"
```

```
    #print data and send to PICAXE
    print serData + " " + serDC
    ser.write(serDC)
    serDataOld = serData
  #if there aren't any visible hands, stop PICAXE activity
  elif len(frame.hands) < 1:
    if not serDataOld == "oo":
      ser.write("o")
      print "oo o"
      serDataOld = "oo"
def main():
  listener = LeapMotionListener()
  controller = Leap.Controller()
  controller.add_listener(listener)
  print(" ")
  print("Press ENTER to quit")
```

```
  print(" ")
  try:
    sys.stdin.readline()
  except KeyboardInterrupt:
    pass
  finally:
    controller.remove_listener(listener)
if __name__ == "__main__":
  main()
```

Lastly, sign up for your free developer account and then download the SDK, the right version for your operating system. Don't forget that PySerial does not work on the Mac, so you won't be able to use the script if this is being done on a Mac. However, the actual Leap Motion and the Leap Motion SDK will work with the Mac.

Step 16 – Now, the Leap SDK is not actually an installer. It is just a collection of the files that you need to make the Leap Motion work with custom scripts. These files will be different depending on which operating system you are using, so copy the right files from the list below for your operating system:

Windows:

32-bit:

- /lib/Leap.py
- /lib/x86/Leap.dll
- /lib/x86/Leap.lib
- /lib/x86/LeapPython.pyd

64-bit:

- /lib/Leap.py
- /lib/x64/Leap.dll
- /lib/x64/Leap.lib
- /lib/x64/LeapPython.pyd

Linux:

32-bit:

- /lib/Leap.py

- /lib/x86/LeapPython.so
- /lib/x86/libLeap.so

64-bit:

- /lib/Leap.py
- /lib/x64/LeapPython.so
- /lib/x86/libLeap.so

Mac:

- /lib/Leap.py
- /lib/LeapPython.so
- /lib/libLeap.dylib

When you have copies of the right files, move them over to the folder that you saved the .py controller to. By now, you should have everything you need to run the Python script properly.

Step 17 – Ok, we have our Leap Motion, we have the correct Leap SDK files, and we have our robot ready to be controlled. There is just one more step that stands between you and being able to move the robot using touchless control. That step is to open up the leappicaxe.py controller and make a change to the serial port settings.

Edit it as you did last time, by changing the portNumber to your own serial ID, and then swap out the commented lines if you are running a Linux system. When you have done that, you can save the edited file and then plug your PICAXE serial cable in. After that plug in the Leap Motion sensor, run the file and, provided there are no error messages, you are ready to go.

Step 18 – It is time to drive our robot. Plug in the USB on the PICAXE serial cable to your PC, using the right port, and plug the 3.5 mm end into the robot. Then plug your Leap Motion sensor into your computer, switch your robot on and then run the Python robot controller. If all goes as it should, you will be able to make the wheels on your robot run by tilting your hands forward and backward.

This version of the robot is easy to control:

- Tilt the left hand forward to drive the left wheel forward.

- Tilt the left hand backward to drive the left wheel backward.

- Tilt the right hand forward to drive the right wheel forward.

- Tilt the right hand backward to drive the right wheel backward.

- Level the right hand to stop the right wheel.

- Level the left hand to stop the left wheel.

Congratulations, you have successfully built yourself a hand-controlled robot!

What if it Doesn't Work?

If your robot didn't work as you expected, have a look at these troubleshooting tips:

- Check that your PICAXE was programmed correctly. If the "blink" program is still up and running, it won't be working very effectively as the brain of your robot.

- Are you batteries new? Old or bad batteries will have a detrimental effect on your robot so either check the ones you have, using a Multimeter, or get some brand new ones.

- Make sure there are no short-circuits or any loose connections anywhere. Check over all of the circuitry and wiring to ensure everything is as it should be.

- Is the Python script running correctly? If you input something incorrectly, when you edited the script or typed it in, nothing is going to work properly. Start again if necessary and make sure you input everything correctly.

- Are you using the right version of Python? It must be version 2.7 or it just won't work.

- Are you using a Mac machine? PySerial does not work on the Mac and, since it is a key part of the project, you need to use it on a Windows machine.

Peter McKinnon

Chapter 10

Making an Autonomous Wall-Climbing Robot

You have now made a simple robot and a robot that knows how to avoid obstacles. Now it is time to make one that can climb up a wall, or any vertical surface, no matter what the material. The robot is equipped with a number of light sensors, which allows it respond to your interactions. This version of the robot will support three different personalities, each of which can be invoked by covering up the top light sensor:

- Red – moves the fastest and will go toward objects.

- Green – moves slower and will avoid objects, turning away.
- Yellow – the slowest of all, will stop in its tracks when it detects another object or motion.

Materials for one robot

- 2 servo motors
- 4 light sensors
- 4 x 2.2K resistors
- 4 x 10K resistors
- 1 x 100-ohm resistor
- 1 Arduino Mini
- 6 x magnetic disks
- 1 x RGB LED
- 1 x lightweight battery
- Hard wire that cannot be easily bent
- Electrical tape

- Wire

- Heat shrink tubing

- Paper or cardboard

- Epoxy or hot glue

Tools

- Hot glue gun

- Soldering iron and solder

- Wire cutters

- Scissors

- Exacto knife

Step 1 – When you purchase servers off the shelf, they are usually fixed motion. However, in order for us to control the movement of our robot, we need to have continuous motion. You will need to modify each of the servos – the physical rotation barriers need to be removed and the potentiometer needs to be hacked to receive a constant signal. The black bit of the motor stops moving after 180 degrees because there are two little plastic knobs that stop it from moving continuously.

- Open up the servo case carefully
- Take the gears apart
- Cut the yellow, green, and red wires that go from the black casing
- Snip the plastic bearings off

That is the first bit, now we need to tell the potentiometer that it will be given a continuous signal:

- Solder one of the 2.2K resistors between the yellow and green wires.
- Solder one of the 2.2K resistors between the yellow and red wires.

A servo will usually rotate in the same direction if it is given the same analog signal. We want to reverse one of them so that they are symmetrical on your robot. We can do this in hardware or code – the hardware way is easier so this is the way we will do it:

Hardware Reverse:

- Cut the blue and red wires that go from the motor board to the motor.

- Solder the blue wire to the red and the red to the blue, criss-crossing them.

Now we need to put the motors back together, so put the wire back in the case as best as you can and put the gears back together. If it doesn't all fit properly, get it as neat as possible and then use electrical tape to tape it all shut.

Step 2 – Next, can attach the wheels to each of the motors. You can use any strong adhesive, including hot glue.

- Cut two pieces of wire the same length, about 1 inch, for each of the wheels.

- Glue the wires to the top gear of each of the servos, making sure that the wire is in the center.

- Glue three of the magnetic disks to the end of each of the wires, again making sure they are in the center.

Step 3 – Now we need to connect the servos onto the Arduino board. Use the Arduino Servo Library and drive the motors using pins 9 and 10. Your setup should look something like this:

- Pin 9 -> Orange wire of Servo 1
- Pin 10 -> Orange wire of Servo 2
- Ground -> Black wires of Servos 1 and 2
- VCC -> Red wires of Servos 1 and 2

Step 4 – Each of the photo resistors needs to be attached to the Arduino. These are the light sensors and there are four of them – one each for the top, left, right, and front of your robot. One of the sensor wires goes to VC and the other goes to the 100-ohm and the 10 resistors. The 10K will connect to ground while the 100-ohm is the input. The green wire, which is the input, on each sensor will go to the analog pins on the Arduino, which are A0, A1, A2, and A3.

Step 5 – Use any RGB LED connected to any of the PWM pins. You could put a resistor between each of the pins and the LED but only do this if your pin cannot take a high current and might burn out.

Step 6 – Now it is time to connect your battery. You can use any light battery between 3 and 4 volts, the lighter it is the better. Attach its power to raw VCC and ground to ground on the Arduino.

Step 7 - Input the code below into your text editor. Each of the sensors is sampled and the robot will move according to which sensor detects your hand and the color – red, green, or yellow.

The following code is a basic program that drives the movement of the motors, based on the input from the light sensors. As I explained earlier, a red robot will move toward objects fast, when the sensor detects some kind of darkness. The green robot will move slower and will move away from the dark areas, while the yellow robots move very slowly and will stop when it detects an object.

These robots are designed to climb a vertical side by using magnetic wheels; the movements that are supported are left, right, and forward. There are a number of different speeds. The light sensors will recalibrate when the robot is rebooted or when the top sensor remains covered over for a period of more than 3 seconds:

```
*/

#include <Servo.h>

// Right and left servos
Servo                                    servo1;
Servo                                    servo2;

// Light Sensors
int    topSensor   =    0;    //700
int  leftSensor  =  1;  ///  Threshold  is  400
int    frontSensor    =    2;    //400
int    rightSensor    =    3;    //300

// Hardcoded thresholds (not used because we auto-calibrate)
int        topThreshhold        =        400;
int        leftThreshhold       =        550;
int        frontThreshhold      =        200;
int        rightThreshhold      =        650;

// Current robot type (red gree or yellow)
int            STATE           =            0;
```

```
// State values
int RED = 0;
int GREEN = 1;
int ORANGE = 2;

// Pins to drive the top tri-color LED
int redPin = 5;
int greenPin = 6;

// Values to hold sensor readings
int front;
int right;
int left;
int top;

// Auto-calibrate light sensor thresholds
void calibrate() {
Serial.println("CALIBRATING");
long int val = 0;
for (int i = 0; i<5; i++) {
val += analogRead(frontSensor);
delay(10);
```

```
}
frontThreshhold      =       (val      /5)      -     80;
val                          =                              0;

for    (int    i    =    0;    i<5;    i++)    {
val       =       val      +     analogRead(topSensor);
Serial.println(analogRead(topSensor));
Serial.println(val);
delay(10);
}

topThreshhold       =       (val       /5)      -200;

val                          =                              0;
for    (int    i    =    0;    i<5;    i++)    {
val              +=            analogRead(rightSensor);
}
rightThreshhold      =      (val     /5)     -    100;
val                          =                              0;
for    (int    i    =    0;    i<5;    i++)    {
val              +=            analogRead(leftSensor);
}
leftThreshhold       =      (val     /5)     -    100;
```

```
// Print threshold values for debug
Serial.print("top:                           ");
Serial.println(topThreshhold);
Serial.print("right:                         ");
Serial.println(rightThreshhold);
Serial.print("left:                          ");
Serial.println(leftThreshhold);
Serial.print("front:                         ");
Serial.println(frontThreshhold);

}

void                                  setup()
{
// turn on pin 13 for debug
pinMode(13,                          OUTPUT);
digitalWrite(13,                       HIGH);
// setup sensor pins
for (int i = 0; i<4; i++) {
pinMode(i,                            INPUT);
}
Serial.begin(9600);
```

```
calibrate();
// generate a random state
STATE = random(0, 3);
setColor(STATE);
}


// MOTOR FUNCTIONS

void turnLeft()
{
Serial.println("LEFT");

start();
delay(20);
for (int i = 0; i<20; i++) {
servo2.write(179);
servo1.write(1);
delay(20);
}
stop();
delay(20);
}
```

```
void turnRight() {
Serial.println("RIGHT");
start();
delay(20);
for (int i = 0; i<20; i++) {

servo2.write(1);
servo1.write(179);
delay(20);
}
stop();
delay(20);
}

void goForward(int del = 20) {
Serial.println("FORWARD");
start();
delay(20);
for (int i = 0; i<20; i++) {
servo1.write(179);
servo2.write(179);
delay(del);
}
```

```
stop();
delay(20);
}

void stop() {
servo1.detach();
servo2.detach();
delay(10);
}

void start() {
servo1.attach(10);
servo2.attach(9);

}

// Set the color of the top tri-color LED based on the current state
void setColor(int color) {
if (color == RED) {
digitalWrite(greenPin, 0);
analogWrite(redPin, 180);
}
```

```
  else if (color == GREEN) {
    digitalWrite(redPin, 0);
    analogWrite(greenPin, 180);
  }
  else if (color == ORANGE) {
    analogWrite(redPin, 100);
    analogWrite(greenPin, 100);
  }
}

// Blink the yellow color (when robot is confused)
void blinkOrange() {
  for (int i = 0; i<5; i++) {
    analogWrite(redPin, 100);
    analogWrite(greenPin, 100);
    delay(300);
    digitalWrite(redPin, 0);
    digitalWrite(greenPin, 0);
    delay(300);
  }

  analogWrite(redPin, 100);
  analogWrite(greenPin, 100);
```

```
}

void loop()
{

top = analogRead(topSensor);
long int time = millis();
while (analogRead(topSensor) < topThreshhold) {
delay(10); // while there is an arm wave from the user don't do anything
}
if ((millis() - time) > 3000) {
// if the sensor was covered for more than 3 seconds, recalibrate
calibrate();
}

// if the top sensor was covered, we change state
if (top < topThreshhold) {
STATE = (STATE+1) %3;
setColor(STATE);
Serial.print("CHANGED STATE: ");
```

```
Serial.println(STATE);
}

// Read the other sensors
right = analogRead(rightSensor);
left = analogRead(leftSensor);
front = analogRead(frontSensor);

if (STATE == RED) {
// go toward objects
if (front < frontThreshhold) {
goForward();
} else if (right < rightThreshhold) {
turnRight();
} else if (left<leftThreshhold) {
turnLeft();
} else {
goForward();
}
}
if (STATE == GREEN) {
// go away from objects
if (front < frontThreshhold) {
```

```
int dir = random(0,2);
if (dir == 0 && right > rightThreshhold) {
turnRight();
} else if (dir == 1 && left > leftThreshhold) {
turnLeft();
}
} else if (right < rightThreshhold) {
if (left > leftThreshhold) {
turnLeft();
} else {
goForward();
}
} else if (left<leftThreshhold) {
if (right > rightThreshhold) {
turnRight();
} else {
goForward();
}
} else {
goForward();
}
delay(200);
}
```

```
if (STATE == ORANGE) {
// only move if there are no hand motions- otherwise blink
int dir = random(0, 3);
if (left<leftThreshhold || right<rightThreshhold || front<leftThreshhold) {
blinkOrange();
} else {
if (dir == 0) {
goForward();
} else if (dir == 1) {
turnRight();
} else if (dir == 2) {
turnLeft();
}
delay(1000);
}
delay(10);
}
}
```

Step 8 - The last step is to make the casing for your robot. As it has been designed to crawl up a vertical wall, we need to make the case as light as we possibly can. Paper or cardboard is ideal but you can also use a lightweight plastic.

Before you do this, make sure that your code works OK and that you can tell which of the servos is on the right and which is on the left.

Hot-glue the servos to the base of your material and then arrange the sensors to the top, front, left, or right of the robot. Make the walls of the body using the same material and then cut out some holes for the sensors and the motor gears to fit into. The top of the case can be whatever light material you want to use.

That is it; your robot is complete and can now be controlled using your hands. Have fun!

All coding in these tutorials is attributed to www.instructables.com

Chapter 11

Cognitive Robotics

Cognitive robotics is a branch of robotics that deals with the intelligent behavior of the robot by providing it with processing architecture that will allow it to process and learn to reason or also learn how to behave as a response to complex goals set by a complex world problem. Cognitive learning of a robot comes under the engineering branch of embodied cognitive science and embodied embedded cognition.

Usually, in order to train a robot there are various algorithms and schemes of a neural network that are applied in order to imitate the structure and the working of a human brain. The traditional form of cognitive learning adapts a method of assuming symbolic coding schemes as a method of understanding the world. It is done by means of translating the world into a set of symbolic representations that has been proven to be problematic if not untenable. Cognitive robotics often deals with the perception and action or even the notion of symbolic representation and hence they are also considered as the core issues of this type of robotics.

Like many great philosophers and psychologists, cognitive robotics starts with the animal cognition and models the development of robotic information processing on it, as opposed to the popular artificial intelligence techniques. This branch of robotics focuses on improving certain capabilities, such as attention allocation, perception processing, anticipation, complex motor coordination, planning, reasoning, and many others. Robotic cognition aims at studying and implementing the behavior of intelligent agents that are present in the physical world (or even in a virtual world, if it is being used in a simulated environment). The goal of this branch is to enable the robot to read and access a situation quickly and respond to it in the real world.

Learning Techniques
Motor Babble

A basic robot learning technique is known as the motor babbling, which often involves correlating a string of pseudo-random motor movements of a robot that is accompanied with a visual or auditory feedback. This feedback may teach the robot to expect its pattern of sensory input and give a corresponding intelligent pattern of motor output. Any required output can be achieved by using a sensory feedback to trigger a signal to inform the motor control. These actions are very similar to a baby in his or her formative years trying to learn talking or walking by reaching for objects around them and learning to produce speech sounds. These concepts are applied to a robot and it should produce a certain response, depending on the input it is being given.

Imitation

When the robot has reached a level where it is able to produce the desired kinetic result and coordinate its motors, it starts to learn by imitation. The robot tries to observe the behavioral performance of a third-party agent and then tries to copy the agent's actions. It is one of the challenges to convert imitation information into a higher-level form of cognitive behavior that can be applied on a complex scene. These patterns are often based on the basic model of embodied animal cognition and imitation is not necessarily required to train a robot to produce high-order complex results.

Knowledge Acquisition

One of the most competitive and complex learning approaches is "autonomous knowledge acquisition," in which the robot is left to observe and perceive its surroundings on its own. There is a database of desired goals and beliefs that is assumed when letting the robot survey its surroundings.

The more direct way of approaching this mode of exploration can be achieved using a "curiosity algorithm," such as the Intelligent Adaptive Curiosity or Category Based Intrinsic Motivation. These algorithms are used to break the sensory input into a finite number of categories and also to find methods of assigning the types of prediction systems (such as the artificial neural network) to each. These prediction systems note the errors and mistakes of the readings and conclusions and stores them for future references. This helps reduce the errors as time passes. This reduction in prediction error is known as learning. The robot then starts to preferentially explore various categories in which learning (also known as reducing prediction error) becomes the fastest.

Other Architectures

There are plenty of platforms on which the architecture of your robot can be based, but most researchers in cognitive robotics have inclined toward architectures such as ACT-R and Soar (cognitive architecture) and lay their bases for cognitive robotics programs. This architecture is highly modular and yet symbolic processing that is mainly used to simulate operator performance and human performance while modeling simplistic data through a symbolized library. The main reason to implement this is to extend these architectures to handle real-world sensory input just as that input unfolds through real time. Now all we need to do is to figure out a way to convert the world surrounding us to simple symbols that can be understood by the robot.

Peter McKinnon

Chapter 12

Cloud Robotics

While the world is swept away with the idea of virtual data storage and the concept of cloud computing, there is a whole new branch dedicated to cloud robotics. Cloud Robotics attempts to inculcate cloud technologies such as cloud storage, cloud computing, and other technologies that revolve around the benefits of a converged infrastructure and shared services for robotics. One of the biggest advantages of connecting the robot to the cloud is that it can be connected to the Internet and can be controlled from any remote location as long as you are also connected to the Internet. The robots also enjoy other benefits, such as powerful computation, communication resources of modern media center in the cloud, storage, and the ability to process and share information from various robots or agents (any other device which is also connected to the cloud server).

This technology enables humans to delegate various tasks through a network remotely while being able to monitor it at any time. The biggest upside of introducing cloud computing to enable your robotic systems is that it can be endowed with powerful capability and yet be able to reduce the costs through cloud technologies. Hence it paves ways for robotics enthusiasts and scientists to build lightweight, smarter robots that have an intelligent "neural network" at the lowest costs. The neural network consists of a data center, knowledge base, deep learning, task planners, environmental models, information processing, environmental models, communication support, and so on.

Components

The main components have been strategically narrowed down to six, which are listed below:

- It offers the robot and its computing software a global library of images, object data, maps, often accompanied with geometry and mechanical properties, an expert system, a knowledge database (such as the semantic web, data centers, etc.)

- Since the robot is connected to the cloud server, it allows massive parallel computation as and when you require it. It is based on sample-based statistical modeling and motion planning, multi-robot collaboration, scheduling, task planning, and coordination of a system.

- Multiple robots connected together are presented with an opportunity to share outcomes, trajectories, dynamic control policies, and robot learning support.

- It is possible to manually share open source code, design and data for programming, experimentation, and hardware construction.

- There is also a facility to invoke human guidance on demand that provides assistance in learning, evaluation, and error recovery.

You can also augment the reaction between human and a robot through multiple ways (such as semantics knowledge or a voice-controlled service like Siri and Cortana).

Applications

Autonomous Mobile robots: Among the best examples of autonomous mobile robots are Google's self-driving cars, which are hugely based on cloud robots. The cloud technology installed in the car gains access to Google's enormous database of maps from the satellite and environmental models (such as Streetview) and finds a method to combine it with the streaming data from the GPS satellites, 3D sensors, and cameras to monitor its own position in terms as small as centimeters. These positions are also coordinated with past and current traffic patterns to avoid collisions. Every car has to obtain information about the roads, environment, driving and its conditions, and it sends information to the Google cloud server, which is used to improve the performance of the other cars.

Cloud Medical Robots: A healthcare cluster (also known as a medical cloud) is a combination of various services and databases such as the disease archive, a patient health management system, electronic medical records, analytics services, practice services, clinic solutions, expert systems, and so on. A robot that is connected to the cloud server will be able to retrieve the information almost immediately and start providing clinical services to patients who may require them and also act as a co- surgery robot and assist the doctors with high-precision surgeries. It helps in establishing a collaboration service by sharing information between doctors, nurses, and care givers about clinic treatment.

Assistive Robots: These are the robots that provide assistance in various situations, ranging from domestic help to assisting in healthcare and life monitoring situations for the elderly or those who require a medical assistance. The cloud system collects and compiles a list of health status of users by transmitting and receiving information with the cloud expert system or doctors to help understand the case of any patient, especially those with chronic diseases. For example, these robots can be employed to provide support to the elderly and prevent them from falling down, or emergency healthcare support in case of an emergency, especially for those with heart disease. The robots will immediately notify the caregivers about the status of the patient in case of a mishap so first aid can be provided on time.

Industrial Robots: "Industry is on the threshold of the fourth industrial revolution. Driven by the Internet, the real and virtual worlds are growing closer and closer together to form the Internet of Things." – quoted from the Germany Industry 4.0 Plan. Cloud-based robot systems are a boon to the manufacturing industries, since they are capable of handling tasks such as threading wires or cables or even aligning gaskets from a professional knowledge base with great precision that cannot be achieved by manual labor. A group of robots can also share information through the cloud server and execute collaborative tasks. On top of that, a customer is able to order a customized product to manufacturing robots since they are directly connected to the online order system. In the world of online shopping and delivery system, a potential paradigm is executed using a cloud server. Once the order is placed online, a warehouse robot is notified of this and it dispatches the item after careful packaging ready for shipping. This package is dispatched to an autonomous car or an autonomous drone that is delivered to the recipient.

Research

RoboEarth: This is a program funded by the European Union's Seventh Framework Program for research, where they fund a lot of technological development projects and a specific department to explore the possibilities of cloud robotics. The main purpose of RoboEarth is to allow one robot to interact and collectively collaborate on tasks and benefit from a shared database of other robots, paving a way for fast advances in machine cognition and behavior. Ultimately, this is a form of subtle and high-tech human-machine interaction. RoboEarth holds a database of cloud robotics infrastructure. Like web browsers, RoboEarth operates on the World Wide Web style of database that stores and collects knowledge generated by humans – and or even robots – in the machine-readable format. Along with the cloud storage, RoboEarth also contains a vast range of knowledge based on software components, maps for navigation (direct connection to the GPS satellite and give object locations, world models, whether conditions, etc.), object recognition models (for example, object models and images), and task knowledge (such as action recipes, manipulation strategy, etc.). The RoboEarth Cloud Search engine extends its support for various mobile, drones, and autonomous vehicles that require a great deal of computation to navigate.

Rapyuta: This is an open source framework for cloud robotics developed by the robotics researcher at ETHZ and loosely based on the RoboEarth engine. Once the robots are connected to Rapyuta's framework, they are able to establish a secured computing environment (or rectangular boxes) that helps them move their heavy computation to cloud. In addition to that, since the computing environments are tightly interconnected with each other, all of them share a high bandwidth connection to the RoboEarth repository.

KnowRob: KnowRob is an extension of RoboEarth that is a knowledge processing system combining knowledge representation and reasoning methods with techniques involved in acquiring knowledge and grounding the knowledge in physical system. It can also serve as a common semantic framework that is used for integrating information from various sources.

RoboBrain: It is a large-scale computational system that is known for learning the publicly available knowledge from Internet resources, computer simulations, and even real-life robot trials. It tries to accumulate and combine everything on robotics into an interconnected and comprehensive database. It is used in applications that include prototyping for robotics research, self-driving cars, and household robots. The goal of the project is self-explanatory from the name: It devises a plan to create a centralized brain that is always online and can tap into the knowledge base at any time. This ongoing project is dominated by Stanford and Cornell Universities and is also supported by the National Science Foundation, The Army Research Office, The Office of Naval Research, Microsoft, Google, Qualcomm, the National Robotics Initiative, and the Alfred P. Sloan Foundation. Their goal is to make the United States of America more competitive in the world economy with the advancements in robotics.

MyRobots: It is a peripheral service that connects the robots and other intelligent agents to the Internet and is regarded as the social network for smart objects and robots, since they are involved in socializing with one another, collaborating, and sharing. This type of sharing helps the robots to benefit by sharing sensor information and giving them insight on their perspective of their current state.

COALAS: It is a project that is funded by the INTERRE IVA France channel – England European Cross-border Co-operation Program, which aims at developing new technologies to help handicapped people through social and technological innovation and also through the users' social and psychological integrity. Its main objective is to produce a cognitively ambient assistive living system that also contains a healthcare cluster in the cloud that is programmed into domestic service robots like humanoids and intelligent wheelchairs with a connection to a cloud server.

ROS: Robot Operating System is used to provide an ecosystem that supports cloud robotics. One of the advantages of ROS is that it is a flexible distributed framework for robot software development. It includes a database that contains tools, libraries, and conventions that are aimed to simplify the tasks, especially those involved in the creation of complex and robust robot behavior across a wide range of robotic platforms. The library for ROS is based entirely on Java implementation and is known as rosjava. This allows android applications to be developed for robots.

Peter McKinnon

Chapter 13

Autonomous Robotics

As the name suggests, an autonomous robot is a robotic device that performs tasks or behaviors with a high degree of autonomy in its decision-making capabilities without the intervention of man. This feature is particularly desirable in fields such as space exploration, wastewater treatment, household maintenance (such as cleaning, washing, etc.), and for delivering goods and services.

Most of the modern factory robots are "autonomous" and abide by strict confines of their direct environment. They understand that not every degree of freedom exists in their surrounding environment and also are intuitive when it comes to defining their workspace inside a factory since it is often chaotic and contains unpredictable variables. It is essential for a robot to know about not only its own exact position and orientation, but also its neighboring robots (in case of big and advanced factories). Even the required task and type of object must be defined. These parameters can vary greatly in a very unpredictable fashion, at least from the robot's point of view.

One of the crucial areas of robotics research is to enable any robot to cope and function in its environment irrespective of where it is, on land, underwater, in the air, underground, or even outer space.

A fully autonomous robot should be able to

- Obtain information about its environment.
- Work for extensive periods of time without the intervention of human beings.
- Remove or move either a part of or all of its body throughout its operating period in the environment without human assistance.

- Avoid situations that are harmful to people, animals, property, and/or itself unless it is a part of the design specification to discard a part of itself.

- A fully functional autonomous robot may also gain new knowledge and learn to adjust to new methods of accomplishing its tasks and adapting to its changing surroundings. Unlike other robotics systems, autonomous robots will require regularly timed maintenance.

Autonomous Navigation
Indoor navigation

A robot tries to associate itself with the behaviors of agents local to the environment and it has to obtain knowledge about where it is and how to navigate from point A to point B. In the 1970s, such navigation required wired guidance on how to move about, but this progressed as the 2000s approached. In the early 2000s, a concept of beacon-based triangulation was introduced that helped the robot pinpoint its location up to a few feet radius. The current commercial navigational robots navigate autonomously by sensing the natural features around them. The first commercial robots to be created were the Pyxus' HelpMate hospital robot and the CyberMotion guard robots, which were created by robotic pioneers in the 1980s. These robots were actually created with CAD floor plans and the entire makeup was done manually. These are sonar-sensing robots that follow variations to navigate their way through a building. The next generation robots were the MobileRobots' PatrolBot and the autonomous wheelchair that come with various customizable specifications, which was first introduced in the year 2004. It also has the ability to create laser-based maps of any particular location such as a building or a locality and is required to navigate in open areas as well as corridors. Their control system is programmed to change its course or path on the fly if something blocks its way.

In the beginning, the autonomous navigation was based on the planar sensors, such as laser range finders, which can sense only at one level. The contemporary systems available in the market now fuse the information available from different sensors for both localization (position) and navigation. There are devices such as Motivity, which relies on sensors that are placed in different areas depending on the sensors that provide the most reliable data; such devices also have the ability to remap a building or a locality autonomously.

For autonomous robots to climb stairs, it requires highly specialized hardware that is used to navigate inside. Most indoor robots can access the handicapped accessible areas, electronic door, and controlling elevators. With the help of such access control interfaces, the autonomous robots are able to freely navigate indoors. Among the most popular topics for research are autonomous stair-climbing robots and those that can manually open doors.

There has been much advancement in the field of indoor navigating robots. The most current ones are the vacuuming robots that can enter and clean a specified room or even the whole floor. Security robots are employed in banks and museums so that they can cooperatively surround the intruders and cut off the exits. These advances also include the concomitant protection in the robot's internal maps, which are typically allow to define the "forbidden areas" so that the robots are prevented from autonomously entering these regions of the building.

Outdoor Navigation

Outdoor autonomy is most easily achieved when flying, since there are few if any obstacles to navigation. Cruise missiles are among the most dangerous autonomous robots since they are armed and if they are hacked they might explode in the wrong place at the wrong time. Pilotless drone aircraft are mostly used for reconnaissance. There are some unmanned aerial vehicles (UAVs) capable of finishing their entire mission with no human interaction. They might possibly need some intervention while landing and that too is done using remote access to the missile. Some missile drones are capable of automatic safe landing. SpaceX has two autonomous spaceport drone ships at sea, one in the Pacific and the other in the Atlantic. They serve as launch and landing platforms for drones.

- Outdoor autonomy is very difficult for ground vehicles for the following reasons:
    - The available disparities in the surface densities
    - Exigencies in weather

- The three-dimensional terrain

- Instability of the environment that is sensed

Peter McKinnon

Chapter 14

Different Types of Robots

One thing you will want to make sure of when you are selecting a robot to build, is that you know what type of robot you will need for your intended purposes to ensure maximum performance. Robots can be used in many different situations, and they are mainly created and designed to reduce human involvement and improve precision. It is important to know the different types of robots, for the knowledge will give you a deeper insight into what to look for and to be aware of throughout your project. Robots fit into one of two broad categories: industrial and domestic robots.

Industrial robots are more commonly used in factories and help with the manufacture of products with precision to make items such as cars, computers, cell phones, medicine, and even food. Robots can work at increased levels of efficiency to help with operating efficiency in manufacturing workplaces. They support industries that manufacture products that are in higher demand, and there is a need to manufacture a great volume of the product in a relatively short time. Every type of industrial robot has a particular use and is specialized to perform a specific function. For an example, robot arms are commonly used in the assembly lines for cars. They can be used to weld frames and to help apply paint. Industrial robots have an advantage over human workers because they are capable of repeating the exact movements repeatedly with great accuracy. Robotic arms are the most common form of robots that are used today. Recently, agricultural robots have been introduced to help with the farm tasks of harvesting crops and cutting weeds, thus saving farmers from having to hire work hands.

The other category is domestic robots. These types of robots are frequently used to perform household tasks.

They are programmed to perform daily chores such as mowing lawns, vacuuming floors, and many other chores that we could not get done, due to our busy schedules. Domestic robots are meant to serve humans, and make our life simpler on a day-to-day basis. You may have seen them, but vacuum robots are becoming very common because they can operate by themselves after their initial setup. With their sensors, they can move around without running into random objects. Some can work on a schedule while others only require the press of a button to start. There are even robots that are beginning to take the place of pets, which allow for easier cleanup and can be managed much better.

When it comes to choosing the robotic platforms, you will want to keep in mind what you will be using the robot for. The next step is deciding which type of robot you wish to build. Most robots are made through an initial inspiration. This usually comes with what you want it to look like and how you would like for it to function. The most important thing is to think up a function for this robot to do and know how to work on everything, so you can achieve what you want. However, if you are a beginner, you will want to start off by going through many different books and robots. Here are a few examples: land robots, stationary robots, hybrid robots, and aerial robots.

Land robots usually have wheels and are among the most common types of robots. Building one is a great place to start if you are just beginning to move into the robot scene. They do not require much money to build, and they can teach you the fundamentals needed for building more complex machines. While there are a few other examples of land robots, such as humanoid robots, they are of higher detail and complexity and, in the end, will take a high level of skill and a vast amount of time to accomplish. Some robots can take years to build due to all the work that must be put together beforehand.

When it comes to wheeled robots, there are advantages and disadvantages. As mentioned previously, wheeled robots are a great place to start if you are looking to learn the core fundamentals of how to build robots. They are simple in design and easy to construct while being light on your wallet. However, if you plan on constructing something that will use six or more wheels, you should consider using a track robot. Another disadvantage of using wheeled robots is that there is a minimum surface area for them to use. If the wheels are not on the ground, you will risk not being able to use the robot at all. This will end up with slipping that will possibly damage your robot.

Another major type of land robot is one that utilizes tracks instead of wheels. One advantage this kind of robot has is that it does not risk slipping as much as wheeled robots. It can evenly distribute its weight, which will help in navigation. They help with ground clearance, and you can add a bigger drive wheel. With advantages often come disadvantages. When it comes to turning a robot with tracks, there is a possibility that it might end up damaging itself, depending on the type of tracks used.

A kind of land robot that is now becoming more popular is the legged robot. They are great when it comes to moving across areas of uneven surfaces. There are some that use only two legs, but most have four to eight. Even though it is easier for them to get around, creating a legged robot takes a lot of programming and a high level of electronics. The battery that is required to provide enough power so that these robots can function to the highest of capabilities, can make it very expensive to build.

Aerial robots have become more popular recently. While we are now able to take to the skies, there will always be problems. The first is that remote control is required most of the time. Most aerial robots cannot function with tethered controllers, and this can lead to a more expensive and complicated robot to build. These kind of robots are becoming so precise, the military is utilizing them around the world.

Aerial robots are great when it comes to remotely controlling aircraft and for surveillance. It is becoming a huge thing among hobbyists who are looking to investigate robotics. A huge disadvantage of aerial robots is that they are very expensive, and they are very easy to destroy. One wrong calculation can send a robot to the ground, thus killing an important project. Another disadvantage is that they need to be controlled most of the time. They have yet to become a real robot that has autonomous control.

Another type is the aquatic robot. While they are increasing in popularity, they are still very complicated to build because they are being submerged into water. They require a lot of parts, many of which can cause this type of robot to be very expensive to build, due to their waterproof capabilities. A significant advantage to these robots is their ability to discover the depths of the oceans. It is a safe alternative to sending someone down to an area that can be quite dangerous to investigate. As I mentioned, this is a very expensive robot to build, and there is always the risk that these robots could be lost to the ocean. From the deep water pressure, the corrosive nature of the saline water to the marine life, there is always the possibility that these robots will never come back. Plan accordingly.

The last major type is the hybrid robot. Many robots do not fit into the categories that have been mentioned in this book. If you build something that is much more complex, there is a good possibility that your robot will need to have parts that come from many different types of robots. The great thing about robots is that you can custom-build them for the job that you need to get done, thus giving you a great amount of control. However, they tend to be very expensive due to the complexities of the project.

Peter McKinnon

Chapter 15

A Deeper Dive into Robotics

By now, you know what the essential parts are and what is needed to construct robots. Now, let us understand more about the parts needed for the functioning of the robots. One of the most essential parts is an actuator.

Some of you may be wondering what exactly an actuator is. It is a device that transforms energy into the physical motion that lets the robot perform its functions. In robotics, this power is typically derived from electric sources. Most of the present-day actuators produce either a linear or rotational motion. The DC motor is an excellent example of this type. When it comes to selecting the right actuator for your robot, it is imperative that you to educate yourself on what they can do and learn some of the basic fundamentals of mathematics and physics.

One type is the rotational actuator. Rotational actuators convert the electrical energy into a rotating motion of the robot. There are two primary mechanical parameters that will differentiate the different types of actuators. The first is the rotational speed, which is usually measured by revolutions per minute (rpm). The second is the torque or the force that the device can create at a certain distance, which is often expressed as Nxm.

An alternating current (AC) motor, is seldom used in robots because of the source of power required. Most robots are motorized through a DC source, usually from batteries. Also, the electronic part typically uses direct current because it is much easier to utilize the same type of energy supply for the actuators themselves. AC motors are primarily utilized in industrial settings. This is where high torque is in greater demand and where the motor can be connected to an outlet to provide the circuitry that is needed.

The DC Motor is commonly in the shape of a cylinder but may also come in several different sizes and shapes. They also are fitted with output shafts that can rotate at extremely high speeds which can often be between 5,000 and 10,000 rotations per minute. Although DC motors are capable of rotating very fast, they do not have a lot of torque. To decrease the speed and add more torque to these engines, you need to install a gear. To install the motor into your robot, you have to fix the motor body to the frame of the robot. Therefore, motors usually have some mounting holes that are located somewhere on the motor's face. This allows for easy installation. DC motors can rotate one of two ways, clockwise or counter-clockwise. The angular movement of the shaft turning can be measured using encoders and potentiometers.

Another type of the main engine for robots is the geared DC motor. This could be added with a gearbox that will increase the torque of the engine, despite it affecting the overall speed. So let's show an example. If a DC motor is currently rotating at about 5,000 rotations per minute and it puts out a 0.0005 Nxm of torque, by adding in a 123:1 gear, would ultimately reduce the speed by a factor of 123, and this would increase the torque by the same factor, 123. There are many types of commonly used gears that you could use: planetary, spur, or worm gears. You can add an encoder to the shaft if you want to know the rotational speed of your motor.

Then there are the hobby servo motors. These motors are also known as aR/C servo motors. These are actuators that rotate to a particular angular position, and they are more commonly used in very expensive remote-controlled robots. They are often used to control the steering or the flight steering in these machines. However, in today's world they are used for several different applications so, instead of being as expensive as in their older days, they are much more cost-efficient and the variety of these motors has changed dramatically. Most servo motors only have the capability to rotate about 180 degrees. This kind of motor is usually made of electronics, gears, a DC motor, and a potentiometer that can to measure every angle. The latter works with the electronics that are installed in order to mobilize the DC motor and to stop the output shaft at a very specific angle. These servos usually have three wires, to control pulse, ground, and voltage in. A servo robot is a more recently developed servo that can supply both continuous rotations and positional feedback. These engines can rotate counter-clockwise or clockwise.

One last type of motor is the stepper motor. Just as the name implies, this is a motor that is capable of rotating as it follows certain steps and degrees. The number of degrees the shaft can rotate with each step varies depending on various factors. The majority of the stepper motors do not include gears, similar to the DC motor. This also means that the torque is much lower. Fixing gears into a stepper motor will have a similar effect as installing gears into a DC motor.

Now let's step out of the engines and move into the linear actuators. These actuators create linear movements. They have three very distinct mechanical properties. They are: The force, measured in kg or pounds; the speed, measured in m/s or inches; and the minimum and maximum distance that the rod can be moved, which is also known as the stroke and is measured in inches or mm.

To start off, there is the linear DC actuator. A linear DC actuator is usually a DC motor that is attached to a lead screw. This allows it to turn as the motor moves. The lead screw has a traveler that is either forced away or is moved toward the motor, which basically pushes the rotating motion in a linear movement. Some of the DC linear actuators also integrate a linear potentiometer that can give some linear positional feedback.

Then there are solenoids. A solenoids is a coil that is wound up, surrounding the mobile core of the solenoid. After the coil is wound up, the core will be forced away from the magnetic field, thus creating motion in a particular direction. If there are several coils, they will provide movements in several other directions. A solenoid stroke is commonly very tiny, but very fast. The strength of the solenoid depends primarily on the electrical power that is passing through it and the size of the coil.

Then come the hydraulic and the pneumatic actuators. Pneumatic and hydraulic actuators use air or liquid to create some kind of linear movement. These types of actuators can also have lengthy strokes, high force, and high speed. In order to use these actuators, you will need to use an air compressor or fluid. This is turn will make it much harder to use when you are comparing them to standard electrical actuators. These are often used in industrial settings because of the amount of force, high speed, and the overall size of the motors.

The next thing to consider is a muscle wire. Muscle wire is a very specialized type of wire that contracts when electricity runs through it. After the electricity has run its course, and the wire cools down from the vertical heat, it will return to its original length and shape. This is a form of actuator that is not very strong or fast and creates very long strokes. However, it is one of the most frequent and convenient actuators to use if you need to work on projects that require much smaller parts.

Now that you know of all these different types of actuators, which one should you use for your robot? There are many different questions that you need to ask yourself when you are looking for what you need for your robot. As you may already know, you need to know what your robot is going to be intended for and what it should be capable of. You need to know what its goal is in order to know which actuator to install. The first thing you need to do is to take notice of all the new innovations and the technology that are released at an alarming rate. Almost all technology will be outdated within six months. You need to remember that one actuator is more than capable of performing several different tasks and perform in various contexts.

A very broad question you need to ask yourself is whether you will need to have access to a mobilized wheeled robot. Drive motors will need to be able to carry the weight of the entire robot, and it will more than likely need a gear down. The majority of robots are utilizing skid steering while trucks and cars are more likely to use rack-and-piston steering. If you would prefer to use skid steering, a geared DC motor is highly recommended to use for these types of robots that are using wheels or tracks. Several geared motors provide a constant rotation, and can have discretionary positional feedback through the use of optical encoders. Since the rotation is needed to be limited to such a certain angle, you can begin with a hobby servo motor for steering.

The next question you will need to think about is whether or not you should limit your robot's range of motion. If the range can be restricted to 180 degrees and the torque needed is not a major factor, then a hobby servo motor is your best option. Servo motors are commonly available in various sizes and torques and usually come with angular positional feedback. Most of these motors come with a potentiometer, while some specialized motors use optical encoders. The R/C servos are commonly used, due to their popularity, when it comes to building small walking robots.

Now on to the next question: Is your robot going to be lifting or turning heavy loads? When raising a large weight compared to shifting weight on a flat surface, you will need much more power. Torque should be one of your biggest priorities over speed, and it is ideal to use a gearbox with a more powerful DC motor. You can also use a linear DC actuator, as long as it has a high gear ratio. You can then use the actuator system that will prevent the mass of what is being lifted from falling if there is a disruption of power from its source. This can include clamps or worm gears.

Another major question that you need to ask yourself is, do you need the angle of this robot to be very precise? If so, stepper motors put side by side with a motor controller could provide a very accurate angular motion. These types of motors are ideal when compared to servo motors because they can provide a constant rotation. However, there are also several high-end digital servo motors that use optical encoders and can provide you with very high precision.

Next, will you need to achieve movements in a straight line with your robot? This may be simple, but linear actuators are highly recommended when it comes to moving parts and making them move in a straight line. They are available in several different sizes and configurations. For faster movement, you will need to consider solenoids and/or pneumatics if you want high torque. You will be able to use linear DC actuators or hydraulics and, if the movement that you have in mind requires minimum torque, you can use muscle wire.

Chapter 16

How to Go About Using Motor Controllers and Microcontrollers

Typically considered the brain, microcontrollers are responsible for making all the decisions, computations, and communications within the robot. These are the devices that can execute a program, typically through a series of pre-set instructions. To interact with the outside world that we live in, a microcontroller has a series of electrical signal pins, is also known as the connection, that can be turned on or off using the programming functions of what has been installed. These pins are also utilized for reading the electronic signals that are frequently released by the sensors or other types of devices that determine whether they are high or low. Most microcontrollers these days can measure analog voltage signals. Or the signals could come with a complete range of values other than just two already specified that are using analog to digital converters or ADC. By the usage of ADC, a microcontroller can assign a numerical number to an analog voltage that is neither high nor low.

Now, what can microcontrollers do? Numerous complicated moves can be achieved by setting the pins high and low creatively. Nonetheless, when building them, there are algorithms involved such as data processing or complex programs that are not in the range of the microcontroller because of its natural limitations and speed. Here is an example. In order to light a blinker in a car, you will need to program a repeating sequence in which the microcontrollers that have been previously installed can set a pin to high, and then wait for seconds before turning it to low and then waiting a few more seconds before it jumps back to high. The light that is connected to the pin will then be able to blink open-ended until that pin is turned off or switched over to something else. Likewise, the microcontrollers can be used to take control of several other electronic devices that include actuators when they are to be installed into the motor controller via Bluetooth, Wi-Fi interface, storage devices, and many others. Due to this versatility, the microcontrollers can be found in almost all standard everyday products. There are several types of electrical drives in a home appliance that utilize at least one kind of microcontroller.

On the opposite end from microcontrollers are microprocessors, typically found in the central processing units of personal computers. Microcontrollers do not need peripherals such as external RAM or storage devices to operate. For this reason, microcontrollers are much less powerful than microprocessors where computing tasks are involved. Building circuits and even products that are based on microcontrollers is a much easier task and is much more affordable. This is because it requires minimum hardware parts. Now, remember that microcontrollers can output a very low amount of electrical power through the pins. That is why a generic microcontroller is not able to power solenoids, big lights, power electrical motors, and other direct loads. If it were to do this, it would end up causing physical damage to the controller.

Now on to programming microcontrollers: To start off, there is absolutely no reason to shy away from programming your microcontrollers. It is not like the past, when making a blinker took a more comprehensive knowledge of microcontrollers and at least a few dozen lines of code. Programming microcontrollers these days is relatively easy. You can use the IDE (integrate development environment), which commonly uses modern languages, covers full lines of archives that can conceal almost all collective actions, and offers several handy samples that can help you get started.

The next step is to choose the proper microcontroller for the robot that you plan on building, unless you are into BEAM robotics or you plan on controlling your robot through an R/C system or even a tethering system. For beginners, selecting the proper microcontroller can be a difficult task, one that takes into consideration the product range, the applications of the robot, and the specification for what it is intended to do. There are various microcontrollers that are available these days, including those offered by POB Technology, Pololu, BasicX, BasicATOM, Arduino, and Parallax. There is a list of questions you should ask yourself when you're figuring out what you need.

The first general issue is finding out what kind of microcontroller is most commonly used for the type of robot that you intended on building. When it comes to building robots, it is not a popularity contest, but the real fact is that if a microcontroller has a larger community that supports it or it has been used in a similar project, that can make the design phase of your robot much easier. With this information, you can benefit from others experiences and more critical robotic hobbyists. It is very common for hobbyists to be willing to share pictures, codes, instructional videos, and lessons that they have learned or picked up along the way.

Next, you need to understand the specific accessories for the microcontroller that you will be picking out. If your robot has a list of special needs or if there is a particular accessory or component that is imperative for your job or design, you need to make sure that the microcontroller you are selecting is compatible with your robot. Although many accessories and sensors can be directly interfaced with most microcontrollers that are out on the market these days, some accessories are designed to work with very specific microcontrollers.

The question you need to ask yourself is what special features, if any, you will need for your robot. When you are selecting a microcontroller, it should be able to execute all of the special actions that are needed for the robot you intend on creating. It must also be able to function well with the special instructions that it has to perform. There are several features that are common to almost all microcontrollers. Among these are the ability to perform basic mathematical operations, making decisions, and having digital inputs and outputs. However, there are several that may need some additional hardware such as ADC, PWM, and other communication protocol support. You must be aware of some pin counts, memory, and the speed that your robot will require.

Now to move on to motor controllers. Motor controllers are electronic devices that serve as intermediary devices between a microcontroller, a power supply, and the motor. Despite the fact that the microcontroller decides the direction and the speed of the motors, it does not have enough power to drive them forward. On the other hand, the motor controller can supply the current for the needed voltage, but it does not have the ability to decide how fast the motor must turn.

So all in all, the motor controller and the microcontroller must work together to make the motor move. The microcontroller can provide instructions to the motor controller on exactly how to power up the motors through very basic and standard communication methods such as UART and PWM. Also, some motor controllers can be manually regulated by the sum of an analog voltage that is created through a potentiometer. The weight and the size of the motor controller can vary considerably, from a device that is much smaller than the tip of a pen to a massive controller that weigh several kilograms. The weight and the size often have very little effect on the robot, unless you want to build an unmanned aquatic or aerial robot.

Just as there are many different types of actuators, there are also different kinds of motor controllers. A few types are the brushed DC motor controller, the servo motor controller, the brushless DC motor controller, and the stepper motor controller.

So how exactly do you go about selecting a motor controller? You can choose a motor controller after you figure out which type of actuator you want to use for your robot. Along with that, the current that the motor draws will depend on the torque that is provided. A small DC motor will not be able to carry much of a current, but it also cannot release much torque. A bigger motor can provide a much higher torque, but it will need much more current.

Chapter 17

How to Control Your Robot Through the Use of Sensors

If we were to define the perfect attributes of a robot, we should keep a few things in mind. It should be more than capable of gathering in its surroundings and making smart decisions and then, based on those decisions, act accordingly and efficiently execute the actions based on its calculations. This also includes the option for the robot to become semi-autonomous, with the idea that it would be controlled by the humans, but there would be other aspects of what it could do on its own. A great example is a complex aquatic robot. You can control the necessary motions and functions of the robot while an installed processor can measure and react to the currents that are underwater. This will keep the robot in a unique position while preventing it from drifting in any other direction.

A camera that would be installed into the robot would be able to send videos back to the one who is controlling it, while the sensor can track the water pressure, the temperature, and much more. As soon as the communication line between the robot and the human begins to falter, an autonomous program would take control of the robot and instruct it to rise to the surface in a safe condition. When you need to control your robot, you will have to figure out exactly what level of autonomy you desire. You will need to decide if you want to go wireless, tether, or autonomous.

The first way to control your robot is through tethered (direct-wired) control. This is the simplest way to control any type of robot because you will use a controller, which is usually handheld, that it is physically connected to your robot through a cable or tether. There also knobs, joysticks, switches, button, and levers that will allow you to control the robot; these are usually placed on the controller. This is all possible without the need to add on any sophisticated parts or electronics. In this setting, the power source for the controller is usually directly connected to the motors, and you will be able to control all of that within the controller. These types of robots and machines normally do not have any type of artificial intelligence, and they are mentioned as remote-controlled devices as opposed to true robots.

A second way of controlling your robot is through a wired computer control. This is another method that can integrate a microcontroller into the machines that you are building, but you are still going to be using a tether. You attach the microcontroller to your computer ports, which will allow you to take control of the actions of your robot through a keyboard, a keypad, joystick, or another type of device. By adding a microcontroller to your robot, you may also have to do some programming for how your robot should respond to the input. This may take a little more time, but it will give you some great results when it comes to operating your robot.

Another great way to control your robot is through an Ethernet interface. The robot is connected directly to a router and can also be connected and used by mobile robots. When you control your robot through the Internet, it is usually recommended that you have a wireless connection, because that will allow you much greater freedom in using your robot for its intended job.

If you want to eliminate the use of cables altogether, you can move over to wireless infrared by using transmitters and receivers. This is a very great thing to accomplish if you are just getting into the robot hobby. Infrared control, however, does need a direct line of sight. If you and your robot have almost any kind of barrier in between, there is a good chance you will not be able to take control of your robot and make it function correctly. The receiver needs to be able to see the transmitter directly to send the appropriate data. Infrared remote controls can be used to send direct commands to the infrared receivers. These are to be paired up with microcontrollers that can interpret these signals and then take control of the robot's actions.

Another form of controlling your robot is through radio frequency transmission. You can use a remote control unit, which often uses microcontrollers, in the transmitter and the receiver so that data can be transmitted through a radio frequency. The receiver box usually contains a printed circuit board (PCB) that includes a tiny servo motor controller and a receiving unit. The RF communication needs a transmitter that works with the receiver or transceiver. Radio frequency transmission does not need a clean line of site, unlike infrared. So this gives the great advantage that it can go to great distances or have obstacles in the way and still be able to function. Primary radio frequency devices can allow for data transfers between longer distances and there is no range or limit for other types of RF devices.

A more modern method of controlling your robot is through Bluetooth. This is a different kind of radio frequency that follows very specific protocols when it comes to sending and receiving data. A standard Bluetooth range is typically around the 10-meter mark, although there is a distinct advantage of controlling the robot through any Bluetooth-enabled devices. These devices can include PDA's, laptops, and even smartphones. Similar to radio frequency, Bluetooth can provide two-way communication.

As modern technology continues to develop at a very rapid rate, the ability to control robots through wireless technology and the Internet has arisen. If you want to control a robot through Wi-Fi, you must have a wireless router that can be connected to the Internet and you must have a Wi-Fi device installed on the robot. You are also able to use a device which has a TCP/IP address through a wireless router.

Most very complex robots these days are autonomous. With the most recent developments in technology, you can now use a microcontroller to its full capabilities and can program it to respond to any input that is built off sensors. Autonomous control can come in many different types. There is restricted sensor feedback, which is pre-programmed with no feedback from the surrounding environment, and there is also a complex sensor feedback. If the robot has to be guided through autonomous control, it would include many different sensors and code that will allow itself to figure out something and to take the smartest action in any given situation.

There are several sophisticated methods of controlling the currently popular autonomous robots, such as visual and auditory commands. When it comes to auditory commands, the robot will be able to react to the sound of the controllers' voices and listen to their instructions.

Everything from "bend forward" to "move right" can be incorporated. A visual command is when a robot can look at something through its sensors and, based on what it sees, can act and move forward with the pre-coded set of instructions for the current situation. Instructing a robot to turn to a particular side by showing an arrow that is pointed to that side would involve a lot of programming, which could be quite complicated. Some of these things are not quite feasible, and they require a very powerful level of programming and can take hundreds of hours to code and produce.

Unlike humans, robots are not restricted by the five senses that we are currently using. They can go beyond, sight, smell, touch, taste, and sound. A robot can be equipped with many different electromechanical sensors that will help it to understand and discover whatever is in its current surroundings. One of the greatest challenges in producing a robot is making it mimic a natural human's sense. Since it is so complicated and hard to do, most robotic builders and developers use alternatives when it comes to triggers that are based on natural senses.

Peter McKinnon

Chapter 18

How to Assemble Your Robot and Program It Correctly

By now, you should have an excellent idea of what is needed to build a robot. It takes time, skills, parts, which unfortunately tend to cost money, and you need to be able to bring it all together to reach an end result. After all of this, you will get to a stage when you have to design and build the framework to keep all of your components together and this will give your robot its personalized look and shape.

The first thing you should be worried about is constructing the frame. There is no manual for building the frame you need, and it should be done in a way that the project demands. You may want a lightweight frame but, in the end, it might cost you a chunk of change. However, you may want to build something that is robust and able to carry a lot of equipment. This could also end up being tough to build, expensive to put together, or hefty to create. The frame can be a very complicated and building it can be a time-consuming process.

The next phase is choosing the materials to use for your design. There are many different types of materials that you can use. As you try various materials in the construction of, not only your robots, but other types of machines, you will understand the disadvantages and the advantages of each material and shortlist the best options for your robot. There is also a certain degree of flexibility when it comes to choosing the material that you are to use. Each material offers different attributes and limitations. It is best to experiment a little with a few different options and, once you have figured out what you like and think will work best for your robot, you will then be able to go full scale and get your materials all ready.

Next is the core construction with your materials. There are some very basic materials that are good quality and that can be utilized for your frame. There are also some very incredible cheap materials, such as cardboard, that you can use. Some of your ideas might need material that can bend and move at a whim while other materials need to be as tough and impenetrable. You can construct a very simple robot out of reinforced cardboard if it is intended for a class project and it will look a lot nicer when it comes to the size that you want your robot to be. You can decorate and add personal customizations to it to make the exterior look exactly how you want it. That's a lot better than using steel if there is no real reason to use it.

Now, when it comes to building your form, you will probably want some very durable materials that will last through your robot's life. If you are building something that will be in operation for several years, it is best to find something that can stand the test of time a little longer than cardboard. Many types of conventional materials can hold up better, such as metal, wood, and even plastic. You only need to punch some holes into your materials so that you can hook up all the electronics. A very strong piece of wood is usually quite thick and extremely heavy, but you can find a sheet of metal that is just as durable and flexible, and it might fit your needs a little better. While choosing your materials, you need to keep in mind that, depending on the purpose of your robot, you may need to move to a more expensive material.

If you are anywhere near the stage where you are ready to have an outsourced frame brought it, the best option to acquire a precision cut through a third party or a laser jet. However, you have to be very cautious when it comes to ordering a frame from a third party because you need to know the exact measurements of your frame. If something is off or miscalculated, you will either have to send it back or amend it yourself. This can be a costly mistake, especially if the frame you need is on an enormous scale. Many of these different types of third-party companies also provide many other types of services. So, if you know what you want and know exactly how you need it, you may want to look into them because they can bring you closer to your product through methods that you may not be able to access.

While 3D printing is beginning to gain some coverage, it is not always the most reliable option when it comes to getting a frame built. They are usually not the soundest option when it comes to structure, because it has to build in layers, and this can create a weakness in the overall structure. On the other hand, this is a process that can produce very complicated and detailed shapes that cannot be built through other conventional means. So, if you are looking to build a special part that will go on the inside of the structure and not be used externally where it might be damaged, this could be a great way of getting the piece that you need. One 3D component may actually contain many important mounting points for all the electrical and mechanical parts without compromising the weight of the robot. The cost of 3D printing is still rather expensive, but it has become more and more popular over the last 10 years. As a result, prices have started to go down, and they may continue to go down over time.

Now that you have pretty much figured out all of the things you need for your robot, you can begin assembling them. Understand that there will most likely be some unexpected surprises, some good and some bad. If you find yourself in a predicament where you cannot decide what to do, just take a moment a think. If you can make it this far in the process, you should be able to fix your mistake, and hopefully before it costs you a lot of money. You can follow the steps below to build a very simple, structurally reliable, and aesthetic robot frame.

The first is to make sure that you have decided on the material that you will want to be using with your robot.

Then you will gather all of your parts that your robot will need. This will include all electrical and mechanical parts, and you need to make sure you have done the measurements and the calculations for everything that you have. If you run into the situation where something is missing, be sure to look further into it. Sometimes parts come in and they are damaged, or the wrong part arrives, or there are various other problems. Just make sure you have everything ready so that, once you start, you are not stuck at a critical point of building your robot.

Then you need to think and design some ideas for the frame of your robot. Depending on what you are going for, you might want to build something that is a little more elegant and beautiful, even though it may not be the most structurally sound thing. Then there may be a few who do not care about what it looks like because it will be used in some heavy machinery and the toughness and the stability of the robot is what matters more.

One you have figured out the design for your robot, you will want to make sure the structure of your frame will be reliable and that you have all the parts needed to support it. It is always best to check and check again what you are doing. That will that you are going about everything in the required manner.

Next, you will want to draw out every component of the root on paper or on cardboard at a true scale. You will also be able to draw out the parts and put them in a CAD program and print them if you have the opportunity. Making sure that everything will fit and be able to operate where you are laying them out will save you much hassle before you move into the construction phase of this process.

Now, you will want to test your design in CAD and an actual setting by using your paper prototype. This will be the test fitting, making sure every component and connection can work where it is located.

Now you want to make sure that you take all of the measurements again, and when you are 100 percent certain that your design is completely right, you can cut the material for the frame. Be sure always to measure twice and cut only once.

You will want to test-fit every part of your design when it comes to assembling the frame. Do this to make sure that, if you need to make some changes, you can do it earlier on rather than have to tear everything apart later.

From here, you will want to construct the frame by using the appropriate assembling materials. Depending on the material that you are currently using, you could use a variety of things. It could be glue, nails, duct tape, welding, screws, or many other different types of binding tools for your project.

Then you fit all of your parts into the frame and, if you have done everything correctly, you will have a robot frame that is seamless and efficient as you need it to be. From this point, all you have to do is build the robot within the frame and everything will be close to finished.

Now that you have the frame built, it may take some time and delicacy to put all the parts together. If it is a time-consuming process, do not be discouraged. Sometimes the best things in life take a little work to achieve. From everything that you have read up until this point, you should have the parts that you need for your robot, including the electrical parts, which are the actuators, the microcontrollers, and the motor controllers that you need. The next step in the process is to put them all together so they can work in the same fashion. As you get further into the mechanical and electronic aspects of building your robot, you will want to make sure that you are following the manuals and taking heed of what they tell you. Certain parts cannot be used in certain ways, so make sure you are not abusing them because, if one is damaged, you will have to replace it and certain parts can be very expensive.

It has come to the point where you need to attach the motor controller to the motor. A geared DC motor or a linear DC actuator commonly has two wires, one black and the other red. You should attach the black wire to the M-

Terminal of the DC motor controller and the red wire should be connected to the M+ terminal. If you were to connect the wires the other way around, the motor would rotate in the opposite direction. While this may not sound too bad, if your motor is connected to something that was not meant to spin backward, you might destroy a part of your robot. If you are using one of the servo motors, there will be black, red, and yellow wires. A servo controller motor comes with pins that are matched with these wires so you should be able to plug them in directly and be able to get going.

When it comes time to connect your microcontroller to the motor controllers, you need to pay close attention again. Microcontrollers come in many different variations and can communicate through several different ways with the motor controller. You need to make sure that you are referring to the manual so that each of your microcontrollers is used according to the instructions to make the proper connection. Regardless of the method that you choose, the microcontroller and the logic of the used motor controller should share the matching ground reference. This can be done by attaching the GND pins together. In the meantime, logic level shifters will be needed if the devices that are being used and do not share the same logic levels. So be sure to know what you need.

When it comes to attaching the batteries to your microcontroller or a motor controller, details are critical. Most motor controllers that are available today have two screw terminals for batteries; they are labeled with a B- and a B+. Even if the batteries are provided with the controller and connect to the controller with the help of screw terminals, you still need to search for a pairing connector that attaches to the screw terminals. If this is isn't possible, you will need to find another way to link the battery to the motor controller while still unplugging the battery and lining it to the required charger. It is possible that not all the electrical and mechanical components that you have selected for your project need to operate at a single voltage. You may, therefore, need several voltage regulations batteries or circuits.

If you building a robot that has a microcontroller, a DC gear motor, or maybe even a servo motor, it will be very easy to see how a battery may not be able to power every little component directly. It is always best to choose a battery that can directly power as many devices as you need to have in your robot. The battery with the biggest capacity can be connected to the drive motors. For example, if the motors you have selected are rated at a nominal 12 volts, the primary battery will also need to be 12 volts. Then you can use a regulator to make sure to energize a 5-volt microcontroller. MiMH and Li-Po batteries are the most recommended choices when it comes to batteries for medium and small robots. If you select NiMH, you will need much cheaper batteries, and if you use Li-Po, you will need very lightweight batteries. Batteries are very powerful devices and they can easily burn up your circuits and other electrical hardware if you are not careful and if they are incorrectly connected. Always double- and triple-check to make sure that you connected them with the correct polarities and that your device can handle all of the energy that your battery is supplying. If you are at the point of second-guessing, there is never any harm in checking it over again. You do not want to make the mistake of assuming something when it could put your entire project at risk. One little mistake can destroy an entire project and cost a lot of money.

Now it is time to add on the electrical parts to the frame; you will be able to attach all of these components using various methods. Make certain that whatever way you are attaching your electrical parts, you do not use them to conduct electricity. Other usual methods to attaching everything together include hex spacers, screws, double-sided tapes, cable ties, glue, and many others.

Programming your robot may take a lot more of your time. This is usually the very last step in building a robot. There is no point in pre-programming your robot, unless it is a very specific project, without making sure that everything has been built and constructed correctly, everything from the frame and the motor to the electrical parts. If you have followed all of the previous steps as they have been laid out, you should be ready, and the robot should be close to what you wanted. Some dramatic changes include switching out actuators or making an entirely new frame; you will want to do this before you continue further.

Now that your robot is nearly finished, it is still nothing more than a hull until you have programmed it to do exactly what you what it to do. You need to do your research when it comes to this phase. Many different types of resources can be found on the web, in stores, etc. There are several programming languages that you can use to program the microcontroller, which will serve as the mind of your robot. Here are the most commonly used programming languages that you can choose from:

The first is assembly language. This may not be the top language but it is just shy of being a fully fledged computer language, and it can be very difficult to use. It is a great language if you plan to program your robot and ensure complete instruction level control.

Another form of a programming language is BASIC. This is one of the most commonly used languages when it comes to hobbyists. Normally this language is used for educational robots.

Then there is C++, which is a very popular programming language that usually has a top level of functionality while keeping your robot at a good low-level control. However, the is a side variant of C++ called Processing. This enables you to uses simplified code to make programming much easier for you and your robot.

Java is said to be the next step up from C++ because it can offer safety features, but at the loss of some low-level control. There are some producers of microcontrollers, such as Parallax. that are making individual components that can be only used with Java.

One of the best known and most popular languages these days is Python. It is very easy to learn and can be used very quickly and very efficiently when integrating several different types of programs. If you purchased a microcontroller from a well-known producer, there is a very good chance that you can find a book that can instruct you how to use the program properly with the most suggested language for that piece. However, if you would prefer a microcontroller from a smaller time producer, it is crucial that you see what kind of language is compatible with that piece of equipment. This may go without saying, but to look at this early on is most important because it will give you an idea of what you will need to know and how to go about it when it comes to the programming phase of constructing your robot.

Peter McKinnon

Conclusion

With this we have come to the end of this book. We have discussed the history of robotics, different types of robots, and the components needed for building a robot. We have discussed the various types of motors along with their advantages and disadvantages. You can use that knowledge for selecting the right motor for your robot. We have also discussed the different materials used for making robots.

I hope you have found this book useful and I thank you once again for using this book.

Peter McKinnon